THE KING OF LOKOJA

THE KING OF LOKOJA

WILLIAM BALFOUR BAIKIE: THE FORGOTTEN MAN OF AFRICA

WENDELL R. MCCONNAHA

Whittles Publishing

Whittles Publishing Ltd.,
Dunbeath,
Caithness, KW6 6EG,
Scotland, UK

www.whittlespublishing.com

© 2022 Wendell R. McConnaha

ISBN 978-184995-490-7

Printed by Short Run Press Ltd.

Dedicated to Judy,
without whom this book would not have been written

CONTENTS

Foreword xiii

Acknowledgments xv

Introduction xix

Prologue xxv

CHAPTER ONE
SHAPING THE MAN   1

The Orkney Islands 1

Baikies of Tankerness 3

Captain John Baikie and His Family 4

William Balfour Baikie 5

The Naturalist 5

University of Edinburgh 6

Royal Navy 7

Royal Hospital Haslar 8

CHAPTER TWO
BAIKIE'S AFRICA   9

Early Societies 9

European Arrival and Creation of the Middlemen 10

The Arabs and West Africa 11

European Nations Challenge Portugal 12

Human Cargo 12

English Enter Africa and the Slave Trade 14

Forts and Factories 15

Uncharted Interior 16

CHAPTER THREE

EARLY EXPLORATION 18

Saturday's Club 19

Scottish Explorers 21

Mungo Park 22

Bain Hugh Clapperton 24

Richard Lemon Lander 26

CHAPTER FOUR

SETTING THE STAGE 27

John Beecroft 27

Heinrich Barth 28

Macgregor Laird 29

The Model Farm 30

The African Steamship Company 33

CHAPTER FIVE

THE VOYAGE OUT 35

John Dalton and Wilhelm Bleek 35

Cholera Quarantine 38

Into Africa 39

Sierra Leone 41

The Pleiad 43

Cape Coast Castle 45

Death of Beecroft and Meeting Reverend Crowther 46

CHAPTER SIX

THE COMMAND 49

Arrival of the Pleiad 50

Captain Taylor 52

The Quinine Experiment 53

Malaria 54

Contents

Cinchona to Quinine 55

Baikie and Bryson 57

CHAPTER SEVEN

THE KWORA AND TCHADDA 59

The Delta 60

Setting the Stage for God and Trade 62

Taylor is Replaced 64

The Search for Barth 65

Success at Trade 67

CHAPTER EIGHT

THE HERO COMES HOME 69

The Dulti 69

William Carlin 70

Trade and Sickness at the Confluence 71

The *Bacchante* 72

The Toast of London 74

CHAPTER NINE

RETURN TO HASLAR 77

The Men of Haslar 77

The Loss of Richardson 81

Presentations and Disappointment 84

CHAPTER TEN

THE NIGER REVISITED 86

Geographical Society of London 86

Narrative of Betrayal and Treachery 89

Laird and the CMS 93

The Expedition Team is Assembled 94

Misfortune, Delays, and Sickness 97

Trade and Religion 98

CHAPTER ELEVEN

JEBBA ISLAND 100

The *Dayspring* Goes Alone 101

Meeting the Etsu of Nupe 101

Juju Rock and Disaster 102

Communication with the Outside World 105

Rescue and Return 107

CHAPTER TWELVE

LOKOJA 109

Masaba dan Malam 109

Establishing a Trade Center 112

The King of Lokoja 114

He's Gone Native 115

Charting His Own Path 117

Missionary Problems 120

The Government's Consul 121

CHAPTER THIRTEEN

THE JOURNEY TO KANO 124

Return to Bida 125

Arrival in Zaria 126

On to Kano 127

Abdulla's War Camp 129

CHAPTER FOURTEEN

GOING HOME 131

Request to Return Home 132

Sierra Leone 132

The Fallen Son 135

Contents

CHAPTER FIFTEEN

A FRAGILE LEGACY   137

The Forgotten Man 138

A Progressive Colonial 139

Model for Trade 139

Proving the use of quinine 141

Writer and Naturalist 144

A Man of the Scottish Enlightenment 145

Epilogue 153

Bibliography 155

Endnotes 159

FOREWORD

William Balfour Baikie was a surgeon, naturalist, linguist, explorer and government consular who played a key role in opening West Africa. Born in 1825 in the Orkney Islands, he studied medicine at Edinburgh and served as assistant surgeon on several of Her Majesty's ships before being assigned to the Royal Hospital Haslar. During his time at Haslar, he made two voyages exploring the Niger and Benue rivers. Following his second voyage, he elected to remain in Africa for five years developing a model for trade that became the standard for all who would follow.

Dr. Wendell McConnaha describes all of this in his thoroughly researched account of this doctor-naturalist and explorer-trader. It is an important story and I take great pleasure in being asked to write a forward to this book. It was nearly fifteen years ago that I commenced corresponding with the author and indeed eventually conducted him around Haslar on a visit to present both Haslar and the Haslar that Baikie would have known. The visit was to assist Wendell in his long aim to write a biography covering Dr. Baikie's life and his time spent at Royal Haslar.

In the mid-nineteenth century, Haslar was one of the largest Royal Naval Hospitals and the location from which Baikie was sent forth on two expeditions to the West Coast of Africa by one of Haslar's greatest men Sir John Richardson CB MD LLD FRS. Richardson from 1837–1855 and Captain Edward Parry, Governor of Haslar from 1847 to 1852 were both explorers, especially of the Arctic and North Canadian Coasts. It was during Richardson's tenure as Inspector of Hospitals and Fleet that many Royal Naval physicians were to be appointed to ships of discovery with duties as ship's surgeon, naturalists and explorers of the Victorian age.

Sir John Richardson and family lived in a large Georgian house situated adjacent to the main façade of Haslar. Richardson extended the house to include a large dining room with windows that overlooked the front façade and in this dining room he would gather Haslar's physicians for dinner at weekends. Over such lengthy dinners both he and other notables present would discuss suitable placements for surgeon physicians onboard overseas expeditions being undertaken by ships of the Royal Navy. Richardson provided both training and encouragement to those appointed to these missions. Dr. William Balfour Baikie was to be no exception.

By undertaking voyages of discovery Baikie rightly joined the ranks of great naval men such as Lind, Huxley, Clark, Watt-Reid, Armstrong, Menzies, Parry and Richardson who were just some of the royal naval surgeons and explorers who played a part in "Pax Britannica's" global expansion. While Haslar is no longer a Naval Hospital, the name and those who served in the Royal Navy Medical Service as physician – explorers should not be forgotten for they have their place in this island's history.

Eric C Birbeck MVO
Haslar Heritage Group

ACKNOWLEDGMENTS

Assembling the events that comprise the life of William Balfour Baikie was an effort made to satisfy my own curiosity as well as an attempt to honor the man. However, it was in no sense a solo effort. I would be remiss if I were not to give credit to those who made my search possible and to those who each contributed to the final creation. I only hope that I don't inadvertently omit someone who richly deserves credit. Should I have done so, my apologies in advance.

Much of the information on Baikie can be found only in libraries, museums, and offices in London. Although Judy and I spent a great deal of time in both Scotland and England in the preparation of this book, the amount of time necessary was never enough. To that end I turned, as one often does, to family. The Birch and Deane family cousins (Janet, Melissa, Alison, and Tim) and Uncle Burt spent hours searching the files and making copies of records from Kew Gardens, the Colonial Office, the Foreign Office, and the Admiralty. Worthy of special mention is Tim Birch who spent hours of his time on this project when he should have been having fun with his mates. I will be forever in their debt for the tremendous contribution they all made to uncovering Baikie's story.

Early in my research, I contacted the Royal Hospital Haslar to see if anyone could provide information on Baikie's two postings there. I was referred to Eric Birbeck, who was described as the historian of the facility; and someone who "knows more about this institution than anyone." That chance introduction has led to a twenty-year long friendship, to a collaborative research partner, and to my first journal publication on Baikie's study of quinine. I discovered that Eric Birbeck, MVO, had served in the Royal Medical Service for thirty-two years, including overseeing the medical center onboard Her Majesty's *Royal Yacht Britannia* for five years. In 1996 Eric transferred to the civil service at the Royal Hospital Haslar. It was shortly after this that I was invited by Eric to visit him at the hospital. We spent time together and I discovered that no one knows the over 260 years of history associated with the hospital better than Eric. He has since become a founding member of the Haslar Heritage Group and I owe to Eric the information that he provided about Haslar, Baikie's time there, his association with his famous colleagues, and many of the photographs in this book. I also thank him for writing the forward to this book.

Much of Baikie's travel, publications, and history are closely linked to the Royal Geographical Society and their collections have played a huge role in helping me to fill in the gaps. This was especially true when I was researching his personal correspondence, much of which can only be found within the Society's archives. Julie Carrington, Librarian at the Royal Geographical Society, proved extremely helpful in directing my research and providing the needed documentation.

Additional valued sources of information became available through Edinburgh University and the Royal College of Surgeons in Edinburgh. Through them, I learned of Baikie's course of study in becoming a Royal Naval surgeon, his continued interest in natural history, pursued in conjunction with his medical studies, and the honors for which he was singled out by his professors and mentors. I owe a debt of gratitude to Marianne Smith, College Librarian of the Royal College of Surgeons of Edinburgh. Throughout his career, Baikie had a strong working relationship with the Royal Botanic Garden Kew. I was delighted to find that Kew was interested in my book and Kat Harrison, Keri Ross-Jones, and Rachel Gardner have been particularly helpful in providing correspondence and drawings related to Baikie.

To assure the accuracy of this work, I asked experts to vet the sections related to the Baikie family in Kirkwall and experts on African history to verify the accuracy of those sections as well. The experts I turned to were Bryce Wilson and Paul Lovejoy. I had first encountered Bryce Wilson when he oversaw the museums service in Orkney. He helped me by providing me with information from the museum archives at that time. Bryce retired about five years ago as museums manager and is the author of many works on Orkney history, including *Profit Not Loss*, which is about the Baikie family. His critique and his editing of the writing I had done about Orkney and the Baikie family history were a tremendous help to me in setting the record straight regarding how William Balfour Baikie fit into the family history of The Baikies of Tankerness.

Finding an expert on African history proved to be the more difficult task. I have been involved in university education for most of my career; however, African history has not been my focus. I recently retired from the University of Pittsburgh, so I turned to our friend, Juan Manfredi, Vice Provost, for assistance. He in turn contacted Patrick Manning, the Andrew W. Melon Professor of World History and the founding director of the World History Center at the University of Pittsburgh. I began a rather extensive email exchange with Juan and Pat that enabled them to determine the nature of my manuscript and then to locate the best person for the review. It was suggested that I contact Paul Lovejoy.

Paul Lovejoy is a distinguished research professor at York University in Toronto and holds the Canada Research Chair in African Diaspora History. Paul has written extensively on Sokoto, the Muslim caliphate in West Africa, and was also well acquainted with the research documents I had collected on Baikie. He agreed to review the entire manuscript, provided a wealth of additional background information that I was able to incorporate into my work, and made key corrections and clarifications where needed. For this, I thank Bryce, Juan, Pat, and Paul for their extensive efforts in making this work as accurate as it can be.

Mr. R. K. Leslie, former chief librarian, and Ms Alison Fraser, former principal archivist, for the Orkney Library, along with Mr. David Mackie, senior archivist provided many of the images used in the book; Ms Anne Brundle and Ms Joyce Gray, from the Tankerness House Museum; Roy Bridges, Emeritus Professor of History at the University of Aberdeen; and Mrs. Jane Wickenden at the Historic Collections Library at the Institute of Naval Medicine in Gosport each provided valuable support. In addition to the individuals mentioned, I want to acknowledge the staff at the various institutions who helped gather materials and who also have provided assistance. They include those working at Tankerness House, the Orkney Library and Archives, the University of Edinburgh, the Institute of Naval Medicine, the British Museum, Kew Gardens, and the Royal Geographical Society. In each location, I was treated with kindness and the utmost cooperation.

When I believed the manuscript was finally ready to be seen by the outside world. I turned to our good friend Walter Vatter. Walter has spent decades in the New York publishing field and provided valuable advice and support as I prepared to send my work off to the publishing houses for consideration. Its ultimate acceptance is due, I'm sure, to Walter's mentoring.

I also want to also acknowledge all the personnel at Whittles Publishing, especially Keith Whittles and Sue Steven for their tireless work on my behalf. They are consummate professionals and their efforts have resulted in the publication of this book. When a copy editor was assigned to review my manuscript, the gods were smiling. Carole Pearce is thorough, knowledgeable, and an absolute pleasure to work with. Her comments and suggestions were consistently on target and her efforts have significantly improved this book.

I close with two special acknowledgements. I was in Africa because of Steve Danchimah. It was his funding that supported my initial travel to Nigeria. It was our discussion that prompted my original interest and ultimate obsession with Baikie. Steve provided additional print and human resources that he thought would help with my research and writing. His efforts have added immensely to this book. Thank you, Steve, but also thanks to Emeka, Chibuzor, Amadi, Obaid, Chinyere and Zayed.

Finally, my wife Judy was there at the beginning. She traveled to Nigeria with me on four different occasions. We were together in London, Edinburgh, and Kirkwall. When I did research, she was by my side. When I was having difficulty in finding the words, she encouraged me. She read my first through to my final drafts and her suggestions all became part to of the final work. This book is complete because of her. Without her this book would not have happened. Without her, I would not be me.

INTRODUCTION

> I am no advocate for endeavouring to acquire new territory; on the contrary, I think such a proceeding would be prejudicial to our views. We should go into Africa as we would into other foreign countries, as visitors, as traders, or as settlers, doing what we could to improve the race by precept and by practice, but avoiding any violent interference or physical demonstration. If attacked, we should be prepared to defend ourselves, but we should be careful not to give cause for offense. (William Balfour Baikie, 1856)

Although he was a religious man, William Balfour Baikie was not a missionary. He was successful in commerce but he was more concerned with the relationships it created than the profits it earned. He adapted to the African culture, rather than expecting those with whom he lived to change. He saw himself as a friend to Africa when others approached the continent with a desire to alter what they had found.

This approach, treating others as equals, makes sense in today's world. But at the time this quote was written, white Europeans saw black Africans as a separate race; a people without a history prior to their contact with the Portuguese; and, as a group, inferior in every way. Africans were sold into slavery, forced to give up their trade goods at a rate favorable to the white merchants and treated as children incapable of charting their own destiny. Into this mix of hostility and patronization came William Balfour Baikie. His willingness to walk a different path and to speak in a different voice makes him unique for his time; and I believed his story should be told.

My study of William Balfour Baikie has been ongoing for the past 28 years. I decided early on, after discovering that little had been written about the man, to write his biography. Always with the best of intentions, at numerous times over the next two decades, I would gather my research and ever so slowly attempt to mold it into book form. In the ensuing years, work and residence in both in the United States and abroad seemed always to take precedence, pushing the writing well onto the back burner. Upon my retirement from the University of Pittsburgh,

I accepted an administrative position at Tsinghua University in Beijing. Once again living abroad, I found the time to complete a task that I began much earlier; and half-way around the world from where I started it.

My interest in Baikie began while I was working at the University of Chicago. Contacts I made through my administrative duties at the Laboratory School had led to my being asked to assist in the creation of a similar school in Aba, Nigeria. On my first trip to Africa I was accompanied by Kathy and Ann, teachers from the Laboratory School, and by Steve, a Nigerian doctor, who was providing the financial support for the school project. It was 1992 and we had hired a car to take our group from Lagos, in the southwestern corner of Nigeria, to Aba, which is located near the border with Cameroon in the far southeastern corner of the country.

It was July, the rainy season. The trip was long and the journey was slow. As the driver carefully navigated the large potholes and missing sections of road we passed a steady stream of local people walking along the side of the road. Because of the condition of the road we were often moving at the same pace as those on foot. It was hot, very hot and humid. We drove with the windows down and with our heads hanging as far out of the car as far as possible to catch the slight breeze.

A white person traveling in rural Nigeria was, and perhaps still is, enough of an anomaly to draw attention. As we passed through the cities, villages, and countryside we were constantly acknowledged by greetings delivered with enthusiasm and warmth. Eager to learn everything I could, I would ask Steve what each greeting meant. I soon made two discoveries: First, there are over three hundred groups of people within Nigeria, each with a unique language or dialect. This meant that I heard dozens of different greetings. Second, although the greetings each sounded different, they were all saying the same thing, "Hello, white man!"

After recovering from the initial discomfort of being acknowledged by the color of my skin, I began to find the welcomes almost commonplace. People would notice one of us, give a classic double take, and then sing out their salutation. Occasionally the greeting was in English. Usually, it was delivered in the local language. As we were crossing the Niger River near Onitsha a young man shouted what sounded like, "Hello, bay kay."

I assumed this was a local word meaning white man, and I was told by Steve that this was true. He said the man was an Igbo and he was saying, "Hello, Beke." However, almost as an aside, he explained that Beke, or Baikie, was the name of a real person. Steve knew of him through his early studies at school. William Balfour Baikie had traveled and lived along the Niger River in the mid-nineteenth century and had been idolized by the people of the area. As a result, all white men who had followed him to this region were called "Beke" by this group of Igbo and this was still the practice over 100 years after Baikie had lived in the area.

That initial visit to Nigeria lasted one month. Upon my return to the United States I began to read everything I could find about the land and the people. From the beginning I found that much of what I read about West Africa in general, and Nigeria in particular, did not correspond to what I had been told or had seen for myself. There were early writings that more nearly matched my understanding of West Africa, like the works of Mary Kingsley; or contemporary descriptions of the continent provided by authors like Blaine Harden. But most of what I read seemed to have been written by someone who had not seen the Nigeria that I had seen.

My initial attempts at research failed to discover any mention of Baikie. His name seemed to be absent from the journals and historical works related to exploration in West Africa. However, works on the early history of Nigeria and the exploration of the Niger River were in plentiful supply. B. I. Obichere, in *Studies in Southern Nigerian History* (1982) describes Nigeria as a country "lying entirely within the tropics, an irregular rhomboid shape that stretches some 700 miles from south to north and 650 miles from east to west, with a total area of over 356,000 square miles." It is the most populous country in Africa.[1]

Nigeria takes its name from the great Niger river system that is its most dominant physical feature. The Niger and Benue rivers meet in the center of the country and flow in a single channel to the Atlantic Ocean. This "Y" formed by the rivers divides the country neatly into three sections. The Y is also a prominent symbol on the country's national flag. A wide variety of ethnic groups reside within each of the three sections. But generally, in the north the predominant group is the Hausa, in the west the Yoruba, and in the east the Igbo. The people within Nigeria's borders make up the largest single country on the West African coast.

Nigeria, like most of the West Coast of Africa had been "discovered" by the outside world in a familiar pattern. First came the sailors who were attempting to find the route to India. These were the Portuguese, Spanish, Dutch, and finally English sailors, who "explored" West Africa without ever leaving the protection of the coast. Then came individuals like Mungo Park, Hugh Clapperton, and the Lander brothers, who used the overland trails and rivers to travel within the continent. Next came the missionaries like Samuel Crowther, then the merchants like Macgregor Laird and George Taubman Goldie. Finally, the soldiers arrived, to fight, to subdue, and then to rule. In Nigeria, it was Fredrick Lugard who led the military conquest of Africa for Britain. Thus, footholds in Nigeria were established by the explorers in the name of their own country, then these positions were expanded using trade and religion, and then ultimately these loose territorial claims were solidified by the gun.

My first attempt at understanding Nigeria was to read journal summaries or the diaries of the travels of the early explorers and travelers at the end of their respective journeys. Most of these men (and in rare cases, women like Mary Slessor) kept journals and published narratives about their exploits. This was the age of Queen Victoria and imperialism and the telling of the story became as important as the original act. Britain, and indeed the whole of Europe, were hungry for news of the expanding Empire. European readers found the exploration of rivers especially interesting, and the discovery of the source of the Nile or the directional flow of the Niger River, became the stuff that sold newspapers as well as the explorers' books of discovery.

I next moved on to books written by European authors who were contemporaries of the early explorer, missionary, or trader and whose primary motive for writing seemed to be to justify what their colleague had done. This literature is as varied as the individuals who initiated these efforts. Yet, with a few exceptions, the records are built uniformly in terms of a single dominating attitude: they are the journals of men and women who look at Africa resolutely from the outside. The common thread in most of these published works was self-glorification and propagation of the story of the Empire. That is, their understanding that everything the individuals who were the focus of these books had done had been in the best interest of Britain, the Empire, and even the African people.

Thus, Thomas Ricks (2018) in his book about George Orwell states that "even into the 1930s, it was common in English culture to portray the empire as a force for good, transmitting education, trade and the rule of law into the far reaches of Asia and Africa." Orwell wrote sarcastically: "[T]he oppressed are always right and the oppressors are always wrong." He continues, this is "a mistaken theory, but the natural result of being one of the oppressors yourself. I felt that I had to escape not merely from imperialism but from every form of man's dominion over man."[2] Well into the twentieth century Orwell's views were still unique. His belief, which was shared by Baikie, seemed even more out of step during the time Baikie was traveling and living in Africa.

Davidson states that although many books that have been written, with only a few exceptions, all are developed along the same pattern. The anthologies of the European "discovery" of Africa were generally conceived as companion books to the study of Europe. They were usually written during the nineteenth century, and they aim to show why and how Europeans "opened up a continent which they would soon turn into a European possession."[3] Boyles states that it was not difficult for Victorians to perceive that there was a proper, English way to do things, which, if diligently pursued, would inevitably ensure the greatest good for the greatest number of people.[4]

This was also the time of the Social Darwinists. Given that Europeans were superior in both military might and technology, by inference it was thought that Africans must be inferior. Their approach to Africa was one of "natural selection" by which superior nations dominated the "backward races" of the world.[5] Public entertainment, musical hall, popular theater, postcards, and other ephemera of the Victorian consciousness were steeped in the idea that English domination was both inevitable and a good thing.[6] Similarly, Rudyard Kipling and other popular writers of the time captured the reading public's attention with fictionalized accounts of English adventurers in exotic African and Asian settings. Contemporary history textbooks stressed similar themes in schools by emphasizing the morality and courage of the Empire.

Thus, Sir William Geary in *Nigeria Under British Rule* says, "In this book it is my business to tell you the history of Nigeria under British rule, how England has discharged her duty to her black subjects. I say at once that our rule has been a benefit to the West African." He discusses his own adventures in West Africa from his arrival in 1894 and assesses the strengths and weaknesses of the various peoples he encountered; noting, for example: "The Yoruba are the most intelligent of West African races, and they have accepted generally British rule with welcome acquiescence and glory in their British citizenship."[7] So, at least in Geary's view, it seems that the intelligence of the various African peoples was based on how readily they had accepted colonialism, and those responsible for the change had done whatever they felt was necessary and proper as their superiors.

A second factor of these early writings was a generally accepted view that Africa did not exist before the Europeans entered the continent, and that it could not have developed without this incursion. In their 1988 work Oliver and Fage note that, throughout much early history, Africa was not backward; indeed, compared to the rest of the world it "was in the lead."[8] However, this idea would not be introduced into European thought until much later. Sir Reginald Coupland, writing in 1928, stated that these incursions and the colonization that ultimately followed were required for a people "that until the middle of the nineteenth century had no history." African history was not a fit subject for scholarly study because no history had existed until the arrival of the Europeans on the scene.

During my initial review of the literature, the search for Baikie was always uppermost in my mind. I would occasionally see a mention of Baikie, but there was never a whole book, chapter, or even a major reference on him. When he was alluded to it was often simply by name, with a sentence or two about his initial voyage on the Niger River. At the end of two years of research I was no closer to discovering the man I sought than I had been during that mid-July ride in Nigeria. The curiosity roused in me by that initial Igbo greeting began to evolve into an obsession. Who was this man so fondly remembered by the Nigerians and apparently so totally forgotten by the rest of the world? The search for the answer to this question was to occupy my next twenty years or more.

I decided to concentrate my search on Baikie. I spent hours looking in general history books, scanning tables of contents and indexes to see if his name was included among the citations. What I discovered was that very little had been written by, or about Doctor William Balfour Baikie. However, the snippets of information that I did glean began to evolve into a portrait of an individual of extremely varied interests, pronounced intelligence, great courage, and amazing foresight.

I found that Dr. Baikie had proved that taking quinine prior to journeying into the interior of Africa would prevent malaria. As a result, his voyage of 1854 had been accomplished without a single loss of life due to tropical fever at a time when previous explorations had resulted in the death of up to three-quarters of the European participants. I learned that he was an author and recognized as a noted naturalist whose many discoveries had been featured in universities, museums, and botanical gardens throughout England and Scotland. I discovered that Baikie had developed a lucrative trading post at the confluence of the Niger and Benue rivers and had elected to live alone at this site for over five years. While others were attempting to force their way into Africa using superior strength, he had chosen to live among these people in the hope of being treated as their equal. Baikie opened up trade routes, catalogued over fifty African languages, and translated large portions of the Bible into the Hausa language. Yet, when they listed names of the individuals who were deemed to be important to the history of West Africa in general and Nigeria in particular, major anthologies either listed Baikie as a mere footnote or failed to mention him at all.

Finally, in the Purdue University library I discovered the journal of his 1854 expedition, *Narrative of an Exploring Voyage up the Rivers Kwora and Binue*. The story of his initial journey was written in 1856 within two years of the completion of the expedition. Yet it reads as if it were written yesterday. It is almost a travelogue that captures the love the man had for the land and the love and trust he had for the people he encountered. Reading Baikie's own words, I discovered someone who had moved beyond the realm of self-praise. This writing was about his thoughts, motivations, and emotions as he moved along the waters of the Niger and Benue rivers. Gone was the need to promote his own deeds or the value of his work. It was replaced by a self-depreciating sense of humor and a warning to his home country and sponsor to approach the Africans as equal partners. Unlike the journals of men who looked at Africa only as observers, Baikie writes as one who is an insider.

I was learning what this man was like, but why he seemed to have been forgotten by history remained a mystery. Attempting to answer this question has taken me to Nigeria a half-dozen times. But it has also taken me to London, Portsmouth, Edinburgh, Aberdeen, and the Orkney Islands. I have accessed dispatches from Nigeria to the British Foreign Office, reports to and from the Admiralty, communications with the Colonial Office, and accounts provided to and from the Royal Geographical Society. In order to fill the gaps where his records do not exist, I have read the works of other Nigerian explorers who were of the same era and who had visited the same cities and regions that Baikie had. I have discussed Baikie with contemporary African historians in Scotland, Canada, and the United States, and studied the records and documents related to the extended Baikie family. Most importantly, using his journals, reports, and personal correspondence, I have used Baikie's own words to tell his story.

Baikie's period of contact in Nigeria lasted only ten years. Yet, one hundred and fifty years after his brief sojourn there, William Balfour Baikie's name is still used to describe all white people who visit the area. Writing on the centenary marking the explorer's death, the Orkney historian Ernest Marwick noted that not only did some of the Igbo use the word *Beke* to mean "white man," but there was a more extended use of the name. A white child was called *nwa Beke* and the white man's country (Britain) was called *ala Beke.*[9] Marwick opens his article by saying, "Although thousands of white men have come and gone, and Nigeria has been since 1960 an independent nation, there are people still in the little towns along the Niger, as well as in places of power and responsibility in that large and interesting country, who remember with cordial respect the name of William Balfour Baikie."

I have attempted for the reader, but more for myself, to define William Balfour Baikie. My goal is to allow those who read this work to get to know Baikie and to walk with him as he forged his way into West African history. To do this, I have looked at his family and his early life in the windswept Orkney Islands and the role those experiences played in who he would become. I have presented a historical summary of the exploration of the Niger River by those who were there before him. I have also gathered together the spare facts of his life, garnered through documents generated at the time, together with the works that others have written about the locations and events where Baikie was also present, and where possible his own words and incorporated them into his story.

This book is an attempt to define the man beyond his accomplishments. But it is also a journey to determine the quirks of fate and the aspects of his personality that have prevented William Balfour Baikie from receiving the recognition that his accomplishments should have provided.

PROLOGUE

As the *H. M. S. Investigator* plowed forward through the gentle seas it began to change direction, steering toward the West African coastline that was just coming into view on the horizon. As the ship drew closer, the hills of the dark green Freetown peninsula could be seen above the lighter tones of the flat coastal mango belt stretching the entire length of the coast of Sierra Leone. On board the English vessel, in addition to the cargo of raw materials collected at the Lokoja trading station there was an unusually large number of passengers. Their primary charge was the person the *Investigator* had been sent to bring back to England, who stood at the rail.

William Balfour Baikie had now been in Africa for nearly seven years. After completing his second exploring voyage, he had elected to stay in Africa. He had established a trading station at Lokoja where the Niger and Benue rivers merged and his efforts had proved to be commercially successful. But illness and fatigue had begun to take their toll. He had recently received word of the death of his mother and that his father was gravely ill. He had asked the British government to send a ship to return him to England and the *Investigator* had been dispatched to bring him home.

When the ship had finally arrived at the confluence, the captain and crew were surprised that Baikie had come onboard with several African youths of varying ages. Baikie had called him his children and indicated they would be traveling on the ship as far as Sierra Leone. The captain had protested, but Baikie had said, "If they do not travel with us, I will not travel at all."[10] The captain was under strict orders from the Colonial Office to bring Baikie back at all costs, so he had grudgingly relented.

When the rescue ship had arrived at Lokoja, Baikie had seemed in relatively good health. But he had been living in Africa for a long time. Climate and disease could take a toll on a younger man and Baikie was nearly forty years old. The voyage to the delta had passed without incident, but when they had reached open waters, Baikie had begun complaining of headaches and fever. The ship's doctor had been treating him, but nothing seemed to bring relief. They were now nearing Sierra Leone and the Captain of the *Investigator* was anxious to unload the children and spend as little time in port as possible. The sooner he could be underway, the sooner he could deliver William Balfour Baikie to those awaiting him in England.

CHAPTER ONE

SHAPING THE MAN

T oday, if one takes the train from Edinburgh and travels north for several hours, the rails lead out of the Scottish Lowlands past villages and farms that have changed little for generations. The train arrives at Inverness, where a short walk between the train depot and the bus station leads to the coach bound for John O'Groats, a small village on the northern Scottish coast.

The road twists northward for hours through the heart of the Scottish Highlands. Lakes and rugged crags emerge from dense evergreen forests. Pastures dotted with hundreds of sheep surround the manor houses and share the landscape with countless castles and ruins. John O'Groats marks the end of the British mainland. Beyond this point the Atlantic Ocean and the North Sea merge and here, dropped into the middle of this windswept seascape are the Orkney Islands. John Tudor in his 1883 guidebook characterizes the isolation of these islands as "the edge of civilization."

THE ORKNEY ISLANDS

The *Pentland Venture* hauls passengers between John O'Groats and Burwick. Today, the Orkney Islands are easy to reach. In William Balfour Baikie's time, small open sailing vessels located at Skarfskerry, about twelve miles west of John O'Groats, provided a ferry and mail service between the mainland and Orkney.

Approximately seventy-three islands make up the present-day Scottish county of Orkney, but many of them are little more than skerries, isolated rocks in the sea. In the entire island group fewer than one-third of the islands are inhabited and the total population is fewer than 20,000 people. As the ferry ploughs forward through the waves, the sky reaches a point between daylight and darkness where it becomes impossible to tell if people still call any of these isolated dots their home.

Today, Orkney is well known for its Nordic heritage. The Norse began to colonize the islands around 800 CE and before long Orkney became a vital link in their western sea routes.[11] Along with the Norse people came names that are not usually associated with parts of Scotland; and the place names and family names can be traced to these early roots. Thus, Orkney families like the Baikies tend to have names that sound more Norse than Scottish.[12] Baikie is one of the oldest family names in Orkney and is the Norse equivalent of Burns.

The ferry lands at Burwick, where the coach for Kirkwall waits. As the coach continues through the gathering darkness the highway becomes quite hilly. For several miles it is evident that the road is climbing toward the center of the island. Then it begins a descent. As the coach rounds a sharp curve, the town of Kirkwall springs into view. The lights are just beginning to come on in the houses and businesses and the setting sun and the calm waters of the harbor beyond silhouette the spire of the St. Magnus Cathedral. The church at the center of the town is this journey's destination and the coach stops beside it.

Currently the county seat, Kirkwall was in Baikie's youth already the capital of the islands (Figure 1.1). The early nineteenth century would have also been the time that the islands were nearing their peak population of 32,000 inhabitants. With a total coastline of nearly 600 miles, in the nineteenth century various areas of Orkney became prime locations for smuggling liquor into Scotland and on to England. The islands were a major herring port, a point of debarkation for the Arctic whalers, and the port of call for ships leaving for the Hudson's Bay Company in Canada.

The frigid climate and isolation of Hudson Bay posed no threat or hardship to these Orcadians. It was said: "The Orkneymen are the quietest servants and the best adapted that can be provided." Their finest tribute comes from the American historian Bernard de Voto, who said the Orkneymen working in Canada "pulled the wilderness round them like a cloak and wore its beauty like a crest."[13]

Figure 1.1 Kirkwall Harbor © Orkney Library and Archive

BAIKIES OF TANKERNESS

Although he was not born to greatness, William Balfour Baikie was certainly born on the fringes of a great family. The ancestral home of the Baikies of Tankerness, located on Broad Street in Kirkwall, is now the Orkney Museum. It is described as one of Scotland's finest town houses and is surrounded by extensive gardens. The museum is charged with preserving and portraying the culture and history of the islands, but only a small display in one of the rooms is devoted to the story of native son, William Balfour Baikie.

William's father, brother, and sisters were a cadet branch of the family known as the Baikies of Tankerness. In their case, the cadet branch consists of the patrilineal descendants of the brother of the Laird of the Baikie of Tankerness or his younger sons. Title and property passed to the Laird's firstborn son. The Laird's younger sons, and his nephews—cadets—had little authority and inherited little or no wealth to pass on to their descendants. Cadets were expected to, as Bryce Wilson states, "make their own way." Cadet members were expected to maintain the family's social status but could pursue endeavors of their choosing. Thus, William Balfour Baikie entered the world with a very well-known name but without the grand wealth and authority that accompanied the senior branch of the family.

His ancestor, James Baikie, started the family on its road to wealth and influence. Hossack, in *Kirkwall in the Orkneys* says that James "carried on business which rapidly made him the wealthiest commoner in Orkney."[14] James went on to buy and sell property himself but would also advance loans on mortgages at a rate of ten percent. If the loan was repaid, he made a handsome profit. If the borrower was unable to meet the obligation James collected the property, reselling it at an even higher return. It was during one such deal that the Baikie family became permanently connected with the Tankerness lands.

Hugh Marwick, in *The Baikies of Tankerness,* tells of James signing a contract with a Mr. Grote, thus acquiring the lands that would provide the family its name. It is unclear whether this agreement was an outright purchase or the collection of an unpaid debt. However, with this arrangement a substantial amount of property, collectively called the Tankerness lands, passed from Mr. Grote to James Baikie.

Sheila Wenham indicates that a "ness" is a promontory and that "Tanker" is probably an English derivation of the Danish surname, "Tannskari." She describes the peninsula called Tankerness as "a large and valuable tract of land about eight miles southeast of Kirkwall."[15] From that point forward, James and his descendants were known as the Baikies of Tankerness.

Tankerness House is directly across the street from St. Magnus Cathedral, and originally served as the Cathedral's rectory (Figure 1.2, *see colour section*). At the time of the Reformation, when church lands were confiscated and sold at auction, the house ultimately passed into the possession of one Patrick Smythe. This gentleman apparently was strapped for money and he, too, borrowed from James Baikie. When he was unable to repay the loan, the title to the property passed to James. Thus, the former rectory became Tankerness House and became the ancestral home of the Baikies of Tankerness.[16]

CAPTAIN JOHN BAIKIE AND HIS FAMILY

John Baikie, a nephew of the Laird of Tankerness, was born at Kirkwall in 1787. With no option to inherit property or title he entered the Royal Navy in 1800 at the age of thirteen. After years of service and a series of promotions, John commanded the Royal Naval flagships for admirals Russell and Ferrier and saw a great deal of action during the Napoleonic War. His obituary in *The Orcadian* newspaper states, "fourteen years of continuous war did he see." In the Orkney Islands there were many skippers, a designation given to any man who commanded a ship engaged in trade but only two Orcadian men at that time could claim the title of captain, which was reserved for those who had commanded a ship in battle. John Baikie left the sea in 1814 but his promotion to the rank of commander, with the title of captain, did not reach him in Kirkwall until 1854. He would be referred to as Captain Baikie for the remainder of his life.

Little has been written about the remainder of the family. William Baikie wrote extensively, both to and about his father. But most of what is known about his mother and his siblings is what can be gleaned from reading the obituaries and census records. We know from these records that John Baikie fathered a son while in a relationship with a woman named Isabella Muir. The birth records list "Baikie, Samuel, natural son of Lieut. John Baikie R.N. and Isabella Muir, an unmarried Woman, was born 26th March 1821" (Kirkwall OPR 21/6B 1820–1854, page 311, 1821).

Beyond this single reference there is no further mention of Isabella Muir. She seems to disappear from John Baikie's life. William Baikie never mentions his half-brother Samuel in any of his writings and neither William's nor John's obituaries list Samuel among their survivors. One may wonder about the relationship between William and his half-brother. Samuel took his father's surname and became a successful master builder in Kirkwall, erecting, among other projects, the Kirkwall Town Hall. It is also apparent that he contributed a major extension to the Baikie line. Between 1849 and 1854 he and his wife produced three daughters and a son.[17]

We know that Captain Baikie married Isabella Hutton in 1822, a year after Samuel was born. John was aged thirty-five and Isabella was twenty-four. The first child from this marriage, William Balfour Baikie, was born in 1825. William was followed in birth by Katherine, the eldest daughter, in 1828. She died in 1854 while Baikie was preparing to leave on his initial voyage to Africa, her death being due to complications with the birth of her first child. The baby survived, but by the time of her August baptism, Baikie was already in Africa.

William's brother, John Hutton Baikie, was born in 1830 and spent his life as a businessman and member of the council in Kirkwall. While the laird and his family lived in Tankerness House, Captain Baikie and his family lived nearby in a house not far from the current location of the Royal Bank of Scotland. John Hutton Baikie lived there with his mother and father and never married. He died in Kirkwall in 1870.

Eleanor, the youngest child, was born in 1832 and was the only family member to survive Captain John. She never married and was living with her father at the time of his death in 1875. Baikie's mother died in 1862, during the time of his extended second mission in Africa. He received word of her death over a year after it had occurred. John Baikie died at eighty-eight years of age, "after a prolonged period of suffering" as the result of a fall on the doorstep of his home.[18]

In 1825 the directors of the National Bank of Scotland opened the first banking establishment in Kirkwall, and Captain Baikie was appointed agent. All accounts show that John was greatly respected and was a visionary within the community. In 1854, through his recommendation and financial backing, he assisted in founding *The Orcadian* newspaper. It was also largely through his efforts that hospital facilities were constructed on the island, and Captain Baikie was for many years the manager of the Balfour Hospital.[19]

WILLIAM BALFOUR BAIKIE

The birth and baptismal records at the St. Magnus Cathedral list William Balfour Baikie's date of birth as August 27, 1825. He was born at Kirkwall into a junior branch of the wealthiest and most influential family of the region. So, it was from this rather elevated perch that William Balfour Baikie viewed his life in Kirkwall.

In the upper levels of British society during the nineteenth century, the eldest son would inherit the family lands and business. If other sons were born, they would be expected to find other means to support themselves and to chart their own course. For many this meant a colonial posting. The eldest son would remain in the ancestral home and the sons who followed would go to India or to Africa. However, in the Baikie family this trend was reversed. John, the second son, stayed in the family home and assisted his father in business. William became a man of Africa. What can be imagined is that Baikie's background and upbringing in the Orkney Islands had a unique and profound influence on the adventurer he was to become.

Baikie's early education consisted of enrollment at the grammar school in Kirkwall, followed by private tutoring. The tutoring was not individualized study but was a set of private lessons attended by himself, his brother, a host of male Balfour cousins, and the sons of some of the more prominent members of the Kirkwall community.[20] There is no mention that his half-brother Samuel received the same form of education. In his obituary in *The Orcadian* (1865) William is described by those with whom he attended school as being "distinguished among his fellows by his amiability and kindness of disposition and gave early indication of that power of application and love of knowledge by which he has since rendered himself famous." The *Orkney Herald,* on the same date described him as a boy "who was noted for his studious, hard-working, retiring disposition—a youth who kept himself very much to himself, and who, while giving evidence of possessing good abilities, was especially remarkable for perseverance and indefatigable application."

THE NATURALIST

During his upbringing in Orkney, Baikie also developed his life-long interest as a naturalist. His homeland made this a natural outgrowth of his formal education. The flora and fauna of the islands stimulated the innate curiosity in young Baikie. The relative lack of natural predators and ample feeding grounds for a diversity of species provide the Orkney Islands with an abundance of wildlife. Grey and common seals use the shores of the islands to give birth. Many animals common to the Scottish mainland are not found in Orkney. The lack of large predators has not only allowed small mammals to flourish, it has also assisted the bird population. Migratory routes and year-round nesting grounds provide large numbers of birds including puffin, curlew and whooper swans.[21]

Baikie and his boyhood friend, Robert Heddle, took an active interest in learning more about the region. They spent hours exploring, collecting, and cataloguing the natural wonders of their islands. In 1847, at twenty-two years of age, Baikie wrote "A List of Books and Manuscripts Relating to Orkney and Zetland." In 1848, he wrote "Historia Naturalis Orcadensis: Zoology, Part I. Mammalia and Birds Observed in the Orkney Islands," with Robert Heddle as his co-author. Baikie's close connection to Robert Heddle and his extended family would continue for decades, in fact, to the very end of Baikie's life.[22] [23]

The bleak landscape, harsh climate and short winter days also bred a warm and friendly people who would have spent large parts of their free time gathered around their fires. Orkneyjar, a website dedicated to preserving the folklore and traditions of Orkney, quotes Edwin Muir, the Orkney poet:

> The winter gathered us into one room as it gathered the cattle
> into the stable and the byre; the sky came closer; the lamps were
> lit at three or four in the afternoon, and then the great evening lay
> before us like a world: an evening filled with talk, stories, games,
> music and lamplight.[24]

It seems reasonable that in this environment, Baikie would have honed his ability to hear and relate a story, and reading Baikie's narrative of his first African voyage, the voice of the storyteller emerges.

UNIVERSITY OF EDINBURGH

As idyllic as Baikie's early life in Kirkwall must have been, it lasted only until his sixteenth birthday. At that time, he left Orkney to begin the study of medicine at the University of Edinburgh, where he reconnected with another famous resident of Kirkwall. Professor T. S. Traill was a native Orcadian, a distant relative of Baikie and a noted naturalist in his day. It was under Traill's guidance that Baikie refined his interest in natural history and zoology. At Edinburgh Baikie also studied under the direction of Professor Robert Christison. Dr. Christison held the Chair of Medicine and Therapeutics at the University of Edinburgh and was also Professor of Clinical Medicine.[25] Christison was considered the top toxicologist and medical jurist of his day and was appointed personal physician to Queen Victoria at the time when he was working with Baikie.

Baikie studied medicine from 1842 to 1846. The courses he completed encompassed nearly all the medical classes and practical electives offered by the University of Edinburgh at that time. In addition to classes like clinical medicine, anatomy, and surgery, he was also able to include coursework focused on his interest as a naturalist. These courses included natural history and two courses in botany.[26]

In Baikie's final year of medical school Professor Christison made him clinical clerk at the Royal Infirmary. In Baikie's time, this advanced medical training was a unique opportunity offered to only one resident each year. It is an indication of Baikie's skill in medicine, just as his publications attest to his ability as a naturalist. His coursework was followed by clinical practice and upon taking his degree in 1848, Baikie entered the Royal Navy as an assistant surgeon.

ROYAL NAVY

In Baikie's first three years he served on Her Majesty's ships *Volage, Vanguard, Ceylon, Medusa,* and *Hibernia.* These various seaborne assignments would certainly have appealed to Baikie's interest as a naturalist. His initial posting was to the *H.M.S. Volage,* which had been designated a survey ship shortly before his posting. Being charged with surveying duties along the coast of Syria meant that the *Volage* was frequently in port, as their assigned duties included a great deal of work on shore. With no specific survey duties assigned to him, this left Baikie free to collect plant and animal specimens of interest. He had an excellent eye: everything captured his attention and he found it all interesting. Dozens of examples of his finds are still housed at Kew Gardens and at the British Museum, among other institutions.

The posting to the *H.M.S. Vanguard* gave a brief halt to his collecting efforts. The *Vanguard* was part of the newly formed squadron of evolution. This was a Royal Navy scheme that combined all the line-of-battleships and steamers in England into a single experimental squadron. The purpose was a head-to-head assessment of ships of similar classes in a steam versus sail evaluation of worthiness for the Royal Navy of the future. The comparison tests took place off the coast of southern England and the west coast of Ireland, and it was a project that the public found quite interesting. As an assignment it may have appealed to Baikie as well. However, it clearly would have taken him away from his work as a naturalist.

His posting to the *H.M.S. Ceylon* saw his return to the Mediterranean. The *Ceylon* had recently been converted into a troop ship and used to convey personnel between the various ports within the Mediterranean. This provided an opportunity for Baikie to return to his collecting and the dates and descriptions attached to many of his discoveries match the dates and locations of his time on the *Ceylon.*

Baikie's fourth posting was to the *H.M.S. Medusa,* a packet ship. These vessels were used to transfer personnel, mail, and cargo between ports and within the Mediterranean and were used as shuttles between the shore and larger ships that had to remain at anchor. Again, this afforded him great opportunity for his study of natural history, as referenced in several pieces of Baikie's personal correspondence to his family and friends in Kirkwall.

Finally, in what he must have seen as a favored assignment, Baikie was posted to Malta as the assistant surgeon for the *H.M.S. Hibernia,* the flagship for the entire Mediterranean fleet. These Mediterranean assignments provided Baikie with the opportunity to continue his interest in natural history, but his seaborne experiences soon came to an end. In 1851 Dr. Baikie was appointed assistant surgeon at the Royal Hospital Haslar in Gosport, Hampshire, a position he held until 1854, when he embarked on his first African voyage.

ROYAL HOSPITAL HASLAR

It was this posting to Haslar that formed the turning point in Baikie's life. At the time of Baikie's arrival Sir John Richardson, renowned surgeon, naturalist, and explorer was responsible for the medical operations for the Royal Hospital Haslar. He was Baikie's immediate supervisor, but he also became his friend and mentor, encouraging and supporting Baikie in both his medical and natural history interests.

Because of his past explorations and publications, Richardson was well known and respected within scientific circles. He had gathered together many equally successful explorers among the medical staff of the hospital. In addition, his reputation and sphere of influence had allowed him to collaborate with naturalists beyond the confines of the Royal Navy. These contacts included Augustus Wollaston Franks, who would become the head of the British Museum; William Jackson Hooker, head of Kew Gardens; his son, Joseph Dalton Hooker who would succeed his father in heading Kew; and Joseph Dalton Hooker's closest friend, Charles Darwin. These key members of the British scientific community became friends and colleagues of Baikie as well.

CHAPTER TWO

BAIKIE'S AFRICA

When Baikie was assigned to the Royal Hospital Haslar he had no idea that he would soon be headed into Africa (Figure 2.1). In 1850 the entire continent was generally unknown, and the Niger River had first been traveled by Scottish explorer Mungo Park fewer than fifty years earlier. To put Baikie's accomplishments into perspective, one must first understand the penetration by outsiders of Africa, West Africa, and the Niger River in greater detail.

EARLY SOCIETIES

J. D. Fage, in *A History of Africa*, maintains that human social life, in the form of settled societies, began about 7000 BCE around the African lakes and in the river valleys. The rivers provided these earliest inhabitants fertile land, natural irrigation, and protection against animals and neighbors. Along these waterways it was possible to identify the first forms of social life, cultivation of crops, and domestication of farm animals, production of pottery, and the early forms of transport.[27]

Figure 2.1 William Balfour Baikie at his graduation from Edinburgh © Orkney Library and Archive

Except for the Nile, no river played a greater role in the establishment of civilization in Africa than the Niger River. Hollett states that the Niger runs for over 2,600 miles, passing through "humid tropical heat, dense forests, then endless wastes of mangrove swamps, where it finally reaches the Atlantic in the Bight of Benin."[28] Sanche de Gramont calls the Niger the cradle of West Africa. He describes it as "a moving path into the heart of the continent, a long, liquid magic wand, that makes fertile the soil it touches."[29]

The Niger–Benue system was the principal artery of commerce and two inter-related systems of trade were practiced. Some traders in the nineteenth century employed large dugout canoes up to forty feet long (12.2m) with a capacity to haul several tons of trade goods. These major traders traveled all the way from the delta to where the Niger and Benue rivers divide. But most merchants followed a pattern of trade using a more elaborate organization. Central markets were established at strategic points along the river, where small-scale merchants traded their goods for commodities from areas further inland. These market areas were considered neutral zones and were generally not impacted upon by wars or disagreements in the vicinity.

This vast river system soon became a conduit for transportation, communication, and trade among the hundreds of prominent groups located along its banks. The early African traders built simple boats and learned how to steer them through the treacherous waters. Going with the current, they soon managed to reach the colossal delta that forms the mouth of the great river. Here, for hundreds of miles the delta is dissected by creeks, streams, and rivers of varying width and length. Through a period of great trial and error, these early indigenous explorers eventually found their way to the mouth of the river and the area of the Atlantic Ocean that would become known as the Gulf of Guinea. Here they would wait in the creeks and branches of the delta to meet those who would eventually arrive by sea.

EUROPEAN ARRIVAL AND CREATION OF THE MIDDLEMEN

The earliest European written description of "inner Africa" was written by an Italian, Antonio Malfante, and dates from 1447. Malfante visited an oasis on the northern edge of the Sahara Desert that was a major staging post for the trans-Saharan caravans traveling from Arabia. In his writing, he describes "innumerable great cities and territories" lying beyond the boundless expanse of the desert. His descriptions are all secondhand as he, and most Europeans to follow him, elected not to make the trek across the Sahara itself.[30]

In the early fifteenth century the Portuguese were searching for a sea route to India so they could participate in the spice trade. As a first step Prince Henry the Navigator launched expeditions to explore the west coast of Africa. Portuguese contact was sporadic until they discovered that African interactions could produce revenue for Portugal through the trade in ivory and gold. By 1471 the gold trade had been opened and Fernao do Po had explored the bights of Biafra and Benin.[31] The island of Fernando Po, named in honor of this explorer, would prove an integral component in Baikie's two explorations of the Niger River basin.

By 1482 the King of Portugal had established the first Portuguese fortification at El Mina, located in present day Ghana. Thus, ten years before Columbus sailed to America, the first permanent European settlement had been established in West Africa. However, Christopher Lloyd states, "It seems paradoxical that Africa was the first continent to be visited by white men in the process of European expansion overseas and yet was the last to be explored.[32]

The Portuguese sailors found that the various river openings were much too shallow for their ocean-going ships. Also, they had no way of knowing which of the watercourses would lead to the interior. This made travel into the interior impossible for the Portuguese and those who followed. The coastal people, aware of this fact, quickly established a middleman trade to keep the business in their own hands. They would travel to the interior, collect the goods to be traded from those who lived farther inland, and then return to the coast to barter with the Portuguese sailors who had remained on or near their ships.

The arrival on the coast of the Europeans stimulated growth and competition among what had previously been small fishing villages, which quickly evolved into several consolidated trading states. Places like Brass, New Calabar, and Bonny became wealthy centers controlled by very powerful rulers. They would soon evolve into the major slaving ports for the Atlantic trade.[33]

THE ARABS AND WEST AFRICA

In 1650 the Arabs had seized Muscat in the Arabian Gulf and had permanently driven out the Portuguese who had occupied the area since 1507. The people of a coastal kingdom that is part of modern-day Kenya and Tanzania asked the Sultan of Muscat to come to East Africa and do the same. The trade routes of the Indian Ocean were easily navigated and the Omani Arabs soon occupied Mombasa, annexed Zanzibar and controlled a coastal kingdom that encompassed most of the East African coast. The Arabs then extended the East African system of regional trade well into the interior. They developed extensive commerce, trading grain, tools, and fabric for ivory, spices, and minerals.[34] Unlike the Europeans, the Arab traders quickly bypassed the coastal middlemen and their incursion into the interior began.

The main trade route ran through the western desert from modern Morocco to the Niger Bend, leading to the development of Timbuktu as a major trading site. The Arabs anticipated that when they began trading with the sub-Saharan people they would encounter unsophisticated individuals surviving as primitive hunters and gatherers. Instead, they found an established trading system far greater and more sophisticated than they had imagined. In fact, the chief reason the trans-Saharan trade grew so quickly in this early Islamic period is that it linked two already flourishing trading systems.

The Arab's Islamic influence spread into the Hausa states through trade and the Islamic scholars who traveled with the trading parties. Additional caravan routes were established across the desert expanse and trading agreements were made with the people beyond the edges of the desert to the south, several of which—the Hausa, Bornu, Songhai and Mandingo among them—were converted to Islam.[35]

Crossing the Sahara took between seventy and ninety days, depending upon the route taken. The process was laborious and capital-intensive. Arab caravans crossed the desert and brought leather, salt, horses, and textiles from the east. They then collected slaves, gold, and ivory as the primary goods to be carried back from places like Kano and Sokoto in modern Nigeria to the markets in Cairo and Bagdad.[36] Ultimately, many of the goods brought from Africa traveled north through the Middle East and beyond to Europe.

Cities and markets, like Timbuktu, Kano, and Sokoto grew to an enormous size and importance. These market towns did not develop along the rivers but at major caravan crossroads or at the terminus of an Arab caravan route. This penetration of West Africa by the Arab caravans had been going on for four hundred years by the time the Portuguese had settled at El Mina and had been in place for seven hundred years before the rest of Europe began to look to the major rivers as the best way to travel within the continent. However, as the activities of the Europeans along the coast increased, trans-Saharan trade declined in intensity. Ships were much better suited, especially to the transportation of slaves, than the grueling overland route of the caravans. However, though it diminished in size, the Arab caravan routes remained intact and continued to operate until the twentieth century.

EUROPEAN NATIONS CHALLENGE PORTUGAL

For two centuries Portugal benefitted from what was known as "the golden centuries of discoveries." This golden age was from the start of the fifteenth century until nearly the end of the sixteenth century, and the Portuguese initially succeeded in keeping West African trade all to themselves. Vasco de Gama had established the route from Portugal all the way to India and the immensely profitable spices of the East made the Portuguese the envy of all Europe; and each rival nation wanted their share.

In 1454, a charter (called a papal bull) was given by Pope Nicholas V to King Afonso V of Portugal. The bull's primary purpose was to forbid other Christian nations from infringing upon the King of Portugal's rights of trade and colonization in these regions. Taylor states the Portuguese continued to hold their prime place along the West African coast for over a century. "They were operating a 'closed shop' in this period of exploration [that was] as effective as that practiced by a modern trade union." This entitlement was maintained until the second half of the sixteenth century when the Pope transferred his support from Portugal to Spain and the Protestant nations of Europe began to question both papal pronouncements.[37]

Once Portuguese domination had been successfully challenged, the Dutch, Spanish, French, and then the English entered the scene in quick succession. This resulted in trading partnerships between the various European kings and the African rulers whom the European seaborne explorers had encountered. In exchange for textiles and metal ware, West African gold, ivory, and spices began to reach Europe by sea instead of across the Sahara and thus avoided passing through the hands of the Arabs. Each of the European countries saw the other Europeans as rivals. Collectively they saw the Arab traders as their common rival. Regardless of the extent of Arab penetration into the continent and disputes over the claimed boundaries along the coast, relationships between each European country and the indigenous populations were quickly established. That relationship was also destined to move from gold and ivory to slavery.

HUMAN CARGO

The Arab traders had long looked to the sub-Saharan people as a product. Slaves had replaced gold and ivory as their primary commodity on the caravan routes years before the Europeans began exploring the Atlantic coast of Africa. Initially, the Europeans believed they had found equals in the West Africans. African and European kings considered themselves to be brother monarchs and addressed each other as such. "Let them go and do business with the king of Timbuktu and Mali," said Giovanni Ramusio, the secretary to the rulers of Venice. "It is no doubt that they will be well received there with their ships and their goods and treated well and granted the favors that they ask."[38]

Initially, therefore, the Africans had little reason to view the Europeans as anything more than profitable business partners, and the coming of Europeans had little effect on the people of West Africa. Pre-industrial Europe's requirements from overseas commerce were essentially luxury goods like silks, drugs, and perfumes, or specialty crops like spices and sugar. These early incursions showed the Europeans were interested only in gathering goods and produce and were not interested in forming trading partnerships in Asia, America or Africa. However, Africa seemed to offer the least opportunity of all.[39]

A study of ships' logs and the writings of the participants of these early European trading efforts finds two interesting features. First, the trade in non-human goods was profitable to both sides and Europeans at first traded only in traditional commercial goods with the local populations. The Portuguese were primarily interested in gold. The French and English wanted gold, pepper, and ivory, in that order of priority. Second, ship's logs and other documents show that none of the Europeans initially had any interest in trading in slaves. The trade in gold from Africa and spices from India was extremely profitable. In comparison with this cargo, slaves were of little value, and interest in them might have been expected to decline. But two factors came into play during the second half of the fifteenth century: Portugal had developed sugar plantations on Madeira and Sao Tome and their gold mines near El Mina were producing nearly half of all the gold that was extracted from Africa at that time. Each of these endeavors required African labor to remain profitable and slave labor was the most profitable labor of all.[40]

Originally, Portugal was the only European country that was involved in the slave trade, and the trade was quite limited, mostly to slaves used in Madeira, Sao Tome and El Mina. However, Columbus's voyage to the America and the West Indies soon made slaves the most profitable cargo of all. During the fifty years after Columbus landed in the Bahamas, the Spanish conquered Mexico and Peru, and Portugal began the colonialization of Brazil. Large deposits of gold, much larger than those in West Africa, were discovered. Huge plantations producing tobacco, indigo, and sugar were developed. Both the working of the mines and the planting and harvesting from the plantations required large quantities of labor. The problem was solved when the Spanish and Portuguese imported the first group of West African slaves to the Americas in 1510.[41]

Like the Arabs before them, the trading of human cargo soon replaced gold and ivory for the Spanish and Portuguese. Davidson asks, "How was it that early European captains and their backers could treat Africans with respect that was due equals, and yet a later world, setting this aside or forgetting it altogether, could regard Africans as inferior?"—a product to be captured, taken miles from their home and sold into a life of slavery.[42] Part of the answer lies in the radical change in labor demands that resulted in a large-scale slave trade. This worked in several ways to change European attitudes toward Africans. Europeans had to justify the trade to themselves by assuming that Africans were inferior, thus persuading themselves that it was part of God's plan for "lesser breeds" to work for the superior races. Thomas Macaulay foresaw the danger that Europeans might consider themselves a superior class. Towards the

African, the British attitude bore the traditional thinking about the innate racial inferiority of black peoples and that still older dictum of Aristotle that some men are by nature free and others are by nature slaves.[43]

Another part of the answer might be found in the extension of Coupland's idea that Africa had no history." With no history of their own, Africans were "children who had failed to grow up." Being "retarded children" they necessarily forfeited any claim to equality of treatment with other people. Possessing no such claim, they could be pressed into service against their will by superior peoples. Even when slavery was abolished the need to "show these people the path to tread" underpinned the ethos of colonial "trusteeship" and was justified as the only approach to Africa that "made sense."[44]

ENGLISH ENTER AFRICA AND THE SLAVE TRADE

The English were much slower to enter the African arena than the Portuguese, Spanish or Dutch. According to Taylor, "There were times in this period when England could have so established her position in Africa as to be as unchallengeable as she was in India." A primary reason for the decision by England to delay their incursion into African was that Britain's eyes were fixed on her Empire in the East. The English were also initially reluctant to trade in slaves. An Englishman, Richard Hakluyt, wrote the famous *Principal Navigations, Voyages, Traffics and Discoveries of the English Nation* in 1597.[45] Although the initial voyages of English slavers had recently begun, he does not mention that slavery was a legitimate form of trade.

Thus, the Englishman, Richard Jobson, offered the earliest description of the Gambia, a river that became the southern starting point for the exploration of the Niger basin for most of the early explorers. On being offered a "coffle" of slaves during his initial visit to West Africa, he remarked, "We are a people who do not deal in such commodities, neither do we buy or sell one another, or any that have our own shape."[46]

It was not until 1562 that the English transported the first slaves to their colonies in the West Indies; and as far as Britain was concerned the Atlantic slave trade did not become a matter of supreme importance until the middle of the seventeenth century. But when the English settlers in Barbados began planting sugar, they set their long-held reservations aside. Sugar was wildly profitable but it is very labor intensive. The planters initially bought Africans from the Spanish, and soon afterwards, from the English. By 1800 one million Europeans lived in the Americas, but over two and one-half million Africans had been carried there by force.[47] England's interest in Africa stagnated for long periods of time and her move from trade in goods to human cargo was delayed well beyond those of her European counterparts. But soon, slaves replaced gold as England's most valuable export.

Because of sugar, in 1773 the value of British imports from Jamaica alone was five times that from the thirteen mainland colonies. British imports from tiny Grenada were worth eight times those from all of Canada.[48] The British islands in the Caribbean became synonymous with sugar plantations and the plantations relied on African slaves. By the end of the eighteenth century, well over three-quarters of all people worldwide were in bondage as either indentured European servants or African slaves;[49] and once established, England's interest remained focused on slavery for the next 250 years.

FORTS AND FACTORIES

When the early Portuguese and later additional European sailors navigated the African coastal waters each country established its own castles and forts. Mannix and Cowley state that during the eighteenth century there were many changes along the coast. During the first part of the century these castles and forts were becoming less important. There was no longer as much need for protection against the indigenous population as most of the coastal people were eager to exchange captives or other goods for European products. However, the fortifications were still useful as depots for trade goods and there were still forty forts in operation at the end of the eighteenth century, including thirty on the Gold Coast alone.[50]

By the time of Baikie's first voyage in 1854, most of the forts had been replaced by a system of factories. A factory was a ship anchored permanently close to shore or a trading station built on the coast itself (Figure 2.2). It was called a factory because the station or anchorage was presided over by an agent called a "factor." There was generally a small residence for the factor and a warehouse to store the European items to be used for trade, surrounded by a stockade. When the commodity being traded was slaves there was also a "barracoon." Mannix and Cowley describe the barracoon as a stockade within the stockade, which resembled a pen for cattle. A long shed ran down the center to protect the slaves from sun and rain. Down the middle of the shed a long chain stretched, fastened to a stake at either end, the male slaves being secured at intervals along the chain. The women and children were usually allowed to run loose within the barracoon.

One would assume that a flourishing market for raw materials would make the job of factor a sought-after position. But *Black Cargoes: A History of the Atlantic Slave Trade 1518–1865* indicates most factors serving in Africa led miserable lives. The factor usually lived alone, unless he had an African mistress. It was also a dangerous life and he needed to be protected by a small guard of local mercenaries armed with muskets. However, it was sickness rather than attacks from the local populations or other traders that posed the greatest risk of death. Some of the factors died on the day of their arrival. Most were dead within two years of their arrival in Africa, which was regarded as the average term of life of Europeans on the Guinea Coast. Some factors died on the edge of the continent where they had come to make their fortune. They died within sight of the ocean, never daring to penetrate the continent further than the perimeter of the coast.[51]

Figure 2.2 Trading ships and factory at Niger Delta
© The Bodley Head and Random House

UNCHARTED INTERIOR

In *The Search for the Niger* Christopher Lloyd states there were four reasons why the Europeans did not penetrate the interior of West Africa. First, Portuguese travel along the coast of Africa to locate a sea route to India and its wealth of spices had been accomplished. The opening of this sea route from Europe to the riches of Asia, and the almost simultaneous discovery of America in the west, turned European minds to other distant lands and they were much too occupied with America and the Far East to bother with the unhospitable land that was Africa. The advent of slavery changed the trade pattern, but slaves could be collected without needing going beyond the coast, and interest in the exploration of the African interior was put on hold until the end of the eighteenth century.

The second aspect preventing exploration of the interior was the fierce entrepreneurial spirit that had developed among the African coastal peoples. Among these coastal people there was an early understanding that they did not have the resources to stop the intrusion of Europeans into their territory. However, they also viewed the commodities that the white man wanted (gold, ivory, then slaves, and later palm oil) as without limit. By expanding their role as the middlemen they could gain great wealth by traveling beyond their own coastal region, collecting or purchasing the products from the people of the interior, and returning to the coast where the Europeans waited to purchase the goods. So, in the beginning, it made sense from a variety of viewpoints—safety, cost, and management—to remain on, or near, the ocean and to allow the coastal groups to complete their orders. The European traders who remained along the coast became known as the "palm oil ruffians,"[52] and they would prove a major problem for Baikie in the future.

Third, no accurate maps or charts existed for much of the continent. Even if the European traders had been able to move beyond what they could see from the coast, Africa was completely unknown. If one were to look at a map of Africa at any given time between 1600 and 1885, the coast would appear as a giant multicolored ruffle bordering the mainland while the center of the continent was black. This was because rivalry among the European powers had created a hopscotch approach in establishing their claims. For example, Portuguese sailors might navigate the coast to the point where they could identify a recognizable landmark. They would then claim all the land to that point in the name of their king. Next an English, Spanish or Dutch ship would sail a little farther and the ship's captain would claim land from the boundary established by the Portuguese to the next identifiable point along the coast in the name of his ruler and country. Thus, the shoreline became a hodgepodge of territories claimed by alternating European countries. These claims were said to extend inland for varying distances, generally without rationale, because no one was aware of what lay beyond a few miles of the coast.

There were attempts to fill in the gaps. From time to time reports of merchants or explorers made their way to geographers, allowing them to draw tentative locations of cities or geographical features. However, these descriptions of what the early traveler or trader had seen were often misread or misinterpreted by early cartographers. Mistakes were made by the original observer and those creating the maps perpetuated the mistakes and added new errors of their own. Many of the early maps had large areas about which no information existed. When that occurred, the map makers often simply made-up mountain ranges, rivers, lakes and cities with no regard as to whether their speculations might be correct. When they could not even hazard a guess as to what topographical feature might be present, they simply inserted animal pictures to fill the space. As Jonathan Swift observed:

> So, Geographers in Afric-Maps
> With Savage-Pictures fill their Gaps,
> And o'er unhabitable Downs
> Place Elephants for want of Towns.[53]

Up to the end of the nineteenth century much of Africa remained the "Dark Continent;" the "Unknown Continent." Its interior was known as the "Heart of Darkness." Thus, the center of a map of Africa was still largely blank at the time Baikie began his stint in Africa. It would remain so for the next three decades.

As late as the early eighteenth century the best map of Africa showed the Niger rising in a great lake close to the Nile. Two great inland oceans were shown to exist. Congoland is shown, but there is no Congo River. Here and there imaginative geographers planted large towns. Timbuktu and Sokoto are shown, but so are cities that did not, had never, and would never exist. The continent is shown to have three distinct divisions, neatly divided with mountain ranges; again, that never were.[54]

The final reason for the European propensity to remain along the coast was that their susceptibility to disease made exploration a deadly endeavor. It was the lethal climate of Africa that most prevented the exploration of the interior. Fever or dysentery killed off most of the early explorers, missionaries, traders, and soldiers assigned to permanent garrison duty. In 1825 the garrison at Jamestown on the Gambia River included 108 men. Four months later, only 21 survived. Before Baikie's travels, malaria threatened to cut short the life of any European who ventured beyond the coast and the Bight of Benin was referred to as the Whiteman's Grave.

> Beware and take care of the Bight of Benin,
> There's one comes out for forty goes in.[55]

CHAPTER THREE

EARLY EXPLORATION

From 1851 until 1854 Baikie continued working as an assistant surgeon at the Royal Hospital Haslar near Portsmouth. With his posting at Haslar, his collecting and work with natural science had almost come to a complete halt. Baikie was a skilled doctor and an able seaman, and his correspondence indicates he missed the travel offered with his onboard assignments and found little to hold his interest within his assigned medical duties. However, in 1854 he was offered a position that one can assume very much captured his interest.

He would be given a leave from his duties at Haslar to allow him to be seconded to a voyage of exploration on the Niger River in West Africa. The Geographical Society of London was to be one of the major sponsors of this expedition. His finds collected during his early years in the Royal Navy were housed in the British Museum and Kew Gardens. He had published two books and various papers on natural history that enhanced his proven record of discovery. Within the confines of the Society, Baikie had already established a reputation as an accomplished naturalist and his skills matched the Society's needs.

He was also a very well-qualified medical officer. Dr. Baikie's orders stated that he was to serve as assistant surgeon and naturalist on the proposed expedition. As assistant surgeon, his medical responsibilities would be minimal. This would allow him ample time to provide "graphic sketches of the regions watered by the lower Niger and Benue rivers."[56] John Beecroft would head the expedition. Dr. J.W.D. Brown would serve as chief surgeon and would be Baikie's immediate supervisor. Lieutenant Lyons MacLeod would pilot the ship and serve as second in command.

To put Baikie's mission into context it is important to understand that he was not the first explorer to be given the task of exploring the Niger River. In fact, explorers had been trying for the previous sixty years to fulfill the mission now assigned to Baikie. The reasons for this interest were twofold: the problem of slavery and the need to expand trade. According to Davidson, the great questions to be resolved at the beginning of the nineteenth century were how England could further reduce the overseas slave trade by attacking it at its source; and to discover trading opportunities beyond the coastal barriers that had been established by the middlemen chiefs.

To solve either issue required an exploration and understanding of areas not yet visited by any of the European powers. The enormous obstacles to the investigation of the African continent included natural barriers, hostile local people, and the ever-present risk of fever and disease. The various river systems seemed to be the most logical approach to exploring a continent so large and so foreboding. Eventually a group of brave pioneers set about to probe, explore, and map the West African interior by attempting to determine the origin, course, and termination of the various rivers, and it all began over a dinner.

SATURDAY'S CLUB

In 1788, Sir Joseph Banks met eleven colleagues for an evening of dining and discussion near the eastern end of Pall Mall in London at a fashionable tavern known as the St. Alban's (Figure 3.1). They called themselves the Saturday's Club. Banks, the president of the club, was on the threshold of being the most important force in English scientific circles. He was a botanist by interest and had taken part in Captain Cook's voyages in search of the Terra Incognita Australis— the mythical continent in the eastern Pacific.[57]

Geographical study was the passion shared by Banks and the other club members. Cook's voyage had just documented the extent of the Pacific Ocean and conducted a brief exploration of the Continent of Australia. During the evening the conversation turned to Africa and the club members reflected that, "as no species of information is more ardently desired, or more generally useful, than that which improves the science of geography; and as the vast Continent of Africa is still to a great measure unexplored, we should form ourselves into an association for promoting the discovery of that quarter of the world." On that evening the Saturday's Club became the African Association, later to become the Geographical Society of London, and ultimately, the Royal Geographical Society. However, this penetration of the Dark Continent did not happen quickly. And there were significant problems related specifically to the exploration of the Niger.

There are many major African rivers that empty their waters into the Atlantic Ocean. These include the Senegal, Gambia, Niger, and Congo rivers. For hundreds of years the people of Africa have made the banks of the River Niger their home. The local people along the river were well aware of its origin, course, and termination. But until the early part of the nineteenth century, the river remained a mystery to Europeans. Part of the confusion over its source and route was due to the length and unpredictable nature of its course. The Niger is over 2600 miles (4200 km) in length. It flows

*Figure 3.1 Joseph Banks ©
Royal Society of London*

through all manner of terrain, from deserts in the north to the mangrove swamps of the southern coast. During what we now know to be its path as it travels toward the Atlantic Ocean at the Bight of Benin, it flows north, east, west, and south.[58]

Another complicating factor was one of language and interpretation. The people found along the banks of the Niger are as varied in culture, ways of living and religion as can be found in any part of Africa. Each group of people had its own unique name for the Niger, hence its multiplicity of names and the confusion it caused Europeans. In his published journal *A Narrative of an Exploring Voyage up the Rivers Kwora and Binue*, Baikie lists twenty-nine local names used to designate the Niger and another nineteen for the Benue, its chief tributary. This would include the names used only by the people along the few hundred miles that he explored. The local people called the upper reaches of the Niger the Joliba and its lower section the Quorra or Kwora. Some called the Benue the Tchadda, from its supposed source in Lake Chad, others called it the Shary or Binue. Where the Niger River finally meets the Atlantic Ocean it is one of over twenty rivers terminating in the delta.[59]

The African Association's desire to explore the Dark Continent reflected the embarrassment felt throughout England and all educated Europe that so vast an area, so close at hand, could still be virtually unknown. Their main goal was to find an inland highway for travel and trade into West Africa. The secretary of the African Association, Henry Beaufoy devised the plan on how the search of interior Africa might proceed. He wrote, "almost the whole of Africa is unvisited and unknown" and that a map of its interior was "still but a wide blank." The Niger River was rumored to be one of the continent's great waterways; however, the "source of the Niger, the places of its rise and termination, even its existence as a separate stream are still unknown." Therefore, the focus of the Association would be to "discover" the interior of Africa by traveling, charting, and mapping the legendary Niger River.

Although exploration was their primary concern, the members of the Association also included some of the titans of British business, and securing a trade advantage for them and for England was also part of their plan. "Gold is there so plentiful," a member of the Association's Committee wrote at their second meeting, "as to adorn the slaves.… If we could get our manufactures into that country, we should soon have gold enough." The Association's work would also be in the interest of Christian learning. And "in the pursuit of these advantages," Beaufoy emphasized, benefits would at the same time be "imparted to nations hitherto consigned to hopelessness and uniform contempt."[60] So, the African Association contained a group of individuals representing political, scientific, and religious interests. Although their goals were not necessarily compatible, they agreed that to a collective end they would concentrate on Africa in general and the Niger River in particular.

The Association initially tackled the problem of the Niger with common sense and good judgment. Elaborate questionnaires were developed and sent to British consuls serving throughout West Africa. The replies were disappointing and simply uncritically repeated the myths and rumors that had developed about the river and its environs over the previous two centuries. Men would need to be sent to explore and provide first-hand information; however, the earliest explorers to be sent out seem to have also been the least competent, compared with those who followed.

Although the goal of these journeys was to focus on the Niger, the initial approach was for the explorers to depart from cities in northern and eastern Africa that were well-established and better known to Europeans. They left from places like Cairo and Tripoli and met with little or no success. Many never returned, and most of the early explorers like Lucas, Ledyard, Horneman, Roentgen, Beckhardt, and Houghton, who were selected by the African Association are now long forgotten. Some of these early explorers died of fever, exposure, and fatigue. Others died at the hands of the local people they encountered. The Atlantic slave trade had been started by white men. The Spanish and Portuguese were still encouraging the coastal groups to sell them this human commodity. The Arabs were collecting as many slaves as ever from the people of the interior. The Africans of the interior and the Arabs both suspected these early explorers of trying to interrupt their primary source of revenue. The coastal people suspected them of trying to take away their livelihood by trading directly with those upriver. None of these groups could conceive of any other reason for wishing to penetrate the continent, and when the men told them they sought knowledge and wished only to discover "when, whence and to where the great river flowed," the Africans and Arabs thought it was but another example of the white man's cunning. It was an atmosphere of fear, competition and jealousy that added so immensely to the difficult task of the first explorers and cost many of them their lives.[61]

SCOTTISH EXPLORERS

After a series of failures, three of the explorers selected by the African Association emerged to secure their place in history. These were Mungo Park, Hugh Clapperton, and Richard Lander. It was these three men who set the stage for Baikie's successes which followed in the middle of the century.

In 1795, the Association agreed to employ a Scotsman named Mungo Park, an Edinburgh University trained surgeon. Park, like many of the "scientific explorers" such as Baikie, was Scottish by birth. Part of their success was that due to the availability of universal public education in Scotland even the poorest Scot had more skills and schooling than their European counterparts. The Renaissance came slowly to Scotland. But when it came, it reached such a flowering that Edinburgh was referred to as "the Athens of the North."[62] Many of the Scottish explorers were also physicians and surgeons and many were educated at Edinburgh University.

Unlike the English universities of the time, the universities in Scotland never became remote ivory towers. By the time Baikie was attending Edinburgh University, Scotland had developed the best educational system in Europe. Elementary schools were set up in every Presbyterian parish, with grammar schools located in the larger towns. Schools were open to all (that is, all boys) and a concerted effort was made in each of these schools to identify the boys with the most ability, regardless of their station or their family's wealth, so that they might go on to the universities.[63]

Despite their small size, Edinburgh and the other universities at Aberdeen, St. Andrews, and Glasgow were international centers of learning and drew students from across Europe as well as from England and Ireland. This was, in part, because non-Protestants could not attend Oxford, Cambridge, or Trinity College in Dublin, while Scottish universities accepted students of all faiths. Lectures at the Scottish universities were conducted in English, not in Latin. Tuition fees in Scotland cost five pounds sterling per year; which was one-tenth of the cost of attending Oxford or Cambridge. These factors meant the graduates in Scotland were much more likely to include students from working-class families, in contrast with the top-heavy landed gentry and aristocratic student bodies in the English programs.[64]

The Scots were also endowed with what Leyburn calls "dourness," which literally means hardness and durability, having the qualities of iron. Those who survive centuries of living in a hard environment, both physical and social, learn how to endure. Such people were much more likely to tolerate the hardships to be encountered on a voyage of exploration, and to accept the minor financial awards offered to these early explorers.

The Scottish medical programs were producing a new type of doctor who combined the roles of a general practitioner, physician, surgeon, and apothecary. The medical student was trained in clinical diagnosis and taught to look at individual objects or parts of nature as components of an overall system. Scottish students were described as being notoriously argumentative and Scottish universities employed the dialectic method because it corresponded with the way the students and their Scottish professors thought.[65]

Unlike their English counterparts, the Scottish university medical curricula places great emphasis on the natural sciences, including botany and zoology. As a result of this new course of study, and unlike their English counterparts, Scottish doctors were concerned with the description, prediction, and understanding of natural phenomena, based on empirical evidence gathered through observation and experimentation. By incorporating an emphasis on clinical diagnosis and system analysis, and by concentrating on the natural sciences, graduates of these programs were ideally suited to scientific exploration and to lead the charge of African exploration.[66]

MUNGO PARK

Mungo means "dearest," and is the alternative name for St. Kentigern, the patron saint of Glasgow. Mungo Park was born into the family of a poor lowland farmer. He was intelligent and ambitious, and his demanding upbringing had made him tough. Upon leaving Edinburgh with his medicine degree, the post of a ship's surgeon was his best option for work, and he secured a position on a ship belonging to the East India Company bound for Sumatra. During the course of this voyage Park continued to indulge his passion for natural history and botanical studies, collecting many rare plants, and upon his return publishing a book about his experiences. When he returned from the voyage he lodged with his brother-in-law, James Dickson. Dickson was an avid gardener and a friend of Joseph Banks, President of the African Association.

The urge to travel had now taken hold of the twenty-three-year-old Park and Dickson recommended him to Banks, who inquired as to his qualifications. Park replied, "to bear fatigue I rely on my youth and the strength of my constitution to preserve me from the effects of the climate." He was proud of the success of his initial voyage and was ambitious to "acquire a greater name than any ever had."[67]

Park's instructions were to sail for Africa and upon his arrival to pass on to the river Niger by a route found most convenient. I should ascertain the course, and, if possible, the rise and termination of that river. I should use my utmost exertions to visit the principal towns or cities in its neighborhood, particularly Timbuctoo and Houssa, and that I should be afterwards at liberty to return to Europe under circumstances that appear to me to be most advisable.[68]

Park was to make two journeys on behalf of the African Association, the first between 1795 and 1797 and the second between 1805 and 1806. Because of the repeated failed attempts to reach the Niger from the east or the north he decided on a new tactic. His approach in each case was to land in West Africa at the mouth of the Gambia River and to proceed eastward, connecting with the Niger near what he believed would be its source. He would then travel with the river to locate its termination.

On the first trip he traveled virtually alone. He encountered many hardships but was also treated with kindness by many whom he encountered. He traveled as far as Segou, in present day Mali. There he observed the Niger, called Joliba by the local Bambara people. "I saw," says Park, "with infinite pleasure the great object of my mission, the long sought-for majestic Niger glittering in the morning sun, as broad as the Thames at Westminster and flowing slowly eastward."[69] At that point, the river indeed flows toward the east. This caused confusion later as it supported the theory that the Niger might be a branch of the Nile or the Congo.

He was exhausted and he had no further goods to trade. He returned by the same route he had traveled earlier, until he reached the mouth of the Gambia. Ships between West Africa and the rest of the world were rare and the only passage he could obtain was going, not to England, but to America. His curious route home took him first to the island of Gorée, in present day Senegal, then to the West Indies. There, Park could secure passage on a mail ship bound for England. He arrived in London on Christmas day 1797. He had been away for two years and seven months.

Park was now the first European to have seen the Niger River and he returned to a hero's welcome. While living in London Park published the narrative of his travels. This publication was extremely well received and remains in print to this day. He then returned to Scotland, where he opened a medical practice and married Alice Anderson, the daughter of a Selkirk surgeon. However, he had difficulty in settling down and in 1800 wrote to Joseph Banks indicating his hope that he "might be of some other use to his country."

Park was to wait several years for his second adventure as the African Association continued to send a series of ill-prepared explorers overland. At last, in 1805 Park was given the commission of captain and placed on a ship bound for Gorée. He was joined by his brother-in-law, Alexander Anderson, a surgeon, and George Scott, a draftsman from Selkirk. This exploration was intended to be much better equipped and manned than his earlier venture.

At this time Gorée was an English garrison where troops belonging to the Royal Africa Corps were stationed. The Napoleonic War was raging and skilled soldiers were in great demand. Those sent to this remote site tended to be the dregs of the English army. Thirty-five soldiers, two sailors, a Lieutenant Martyn, and Isaaco, a Mandingo guide, made up the party. They left to follow the same route that Park had taken on his earlier expedition. Unfortunately, the group was too large and too poorly trained to achieve success and they encountered problems from the very beginning.

As always in African exploration, the major threat was from disease. By the time Park once again reached the Niger, only six soldiers, Martyn, and Isaaco were still alive. At the river Park had a large canoe built, but fate was against the group. Three more soldiers died before the work could be finished. Finally, the remaining members of the party began their voyage down river.

They first floated east, as Park had determined earlier, passing the fabled city of Timbuktu. There the river makes a vast bend and begins traveling south and west toward the Atlantic Ocean. The small party continued down the river as far as Bussa. There the Niger narrows into a fast-moving series of rapids, bounded on each side by steep rock cliffs. This locale had long been an ideal spot for adversaries to intercept those traveling on the river. There, while attempting to navigate the rapids, the group was ambushed and killed by the local villagers. Only Isaaco survived to tell others later of his comrades' fate. Park's grand career ended after having traveled over eight hundred miles on the Niger. He had also verified that it flowed west toward the Atlantic, not east toward the Nile or Indian Ocean. However, he did not live to tell his story and it would be left to others to fill in the final blanks.[70]

BAIN HUGH CLAPPERTON

Except for Park's expedition, the exploration of Africa had been put on hold for most of the Napoleonic War. With the end of the war, however, there was once again an opportunity to employ both naval ships and men. Because it was difficult to advance in the military during peace time, serving as the leader of an expedition, or as a naturalist or ship's surgeon, became popular avenues for advancement for hundreds of members of the Royal Navy who were now looking for work.

Hugh Clapperton was a Scottish naval officer who had seen a great deal of action during the Napoleonic Wars. But in 1817 he had elected to leave the Royal Navy and drifted for several years without much purpose. In 1822, he had just moved to Edinburgh. At that time his friend, Walter Oudney, was appointed to lead an expedition to the Bornu Empire in present-day northern Nigeria, where Oudney was to serve as consul, and he invited Clapperton to accompany him.

The expedition's immediate objectives were to gather scientific, geographical and political information. Kolapo describes Clapperton's inclusion on the expedition as "instructive" and notes that Clapperton and Oudney belonged to a different genre of European observers and recorders. Previous explorers had been dispatched primarily to assess the potential for trade, determine the extent of slavery in the region or consider the potential for Christian conversion. Clapperton "considered his area of expertise and duty on the mission was having the honor of being one of the first explorers to make useful discoveries and provide accurate charting of routes and providing information about the states of peace and war."[71]

Clapperton and Oudney left from Tripoli and traveled toward Murzuk, where they were later joined by Dixon Denham. In early 1823 the party reached Kuka (later called Kukawa) which was then capital of the Bornu Empire and the southernmost point of the trans-Saharan trade route from Tripoli. Oudney's letters of appointment were presented, and the group was well received.

Oudney and Clapperton then proceeded with the extension of their mission, which was to explore the course of the Niger River. Oudney became ill and died at Murmur on the road to Kano. Clapperton buried his friend, then traveled alone to Kano, where he remained for nearly six weeks, partly because by this time he too was suffering from recurring fever. Clapperton then continued his journey to Sokoto, the capital of the Fulani caliphate. There he met the Caliph, Muhammed Bello, who seemed to hold Clapperton in high regard and who expressed a desire to cultivate a friendship with England.

Bello asked if the King of England would send him a physician to reside in Soudan and requested that a consulate be established in Sokoto. The Caliph promised to give England "a place on the coast to build a town," only wishing to "have a road cut" to the trading centers, "should they not be able to navigate the river."[72] Clapperton successfully negotiated a verbal trade agreement between the caliphate and Great Britain. He had every reason to be pleased with his success. He had been promised a commercial relationship with Sokoto and he had been assured by the Caliph that there would be no problem in reaching the sea by traveling down the Niger River. His work was done. Clapperton and Denham survived their time in Africa and returned to England in June 1825.

Clapperton was the first to return with a full and intelligent report of the civilizations of the Sokoto caliphate. He seems to have had a real understanding of the historical process in which he was involved, and to have learned much from his interaction of the caliphate's leaders; especially Muhammad Bello.[73] Upon his return to England Clapperton was promoted to the rank of commander and immediately sent out on another expedition to Africa. The official purpose of this second expedition was to formalize what he had negotiated with Muhammed Bello in 1824. That promised agreement would establish a trading center at the confluence of the Niger and Benue rivers, which the Caliph had described in his earlier offer to Clapperton. This would provide the English with a huge trade advantage, as the caliphate was the largest independent country in Africa. The overarching personal reason for his return to Africa was that Clapperton thought he had solved the puzzle of the Niger River and needed a second expedition to prove his theory.[74]

In November 1825 Clapperton sailed on the *H.M.S. Brazen*. He landed at Badagry in the Bight of Benin, where his party included Richard Lemon Lander, who had been hired as his domestic servant for the voyage and with the expectation that he would remain in this role. In July Clapperton arrived at Kano, intending to continue to Bornu. He then traveled as far as Sokoto but was stopped from going further because the Fulani were now at war. He stayed at the site for months, unable to leave. While he was stranded in Sokoto, he was afflicted by malaria and dysentery and he died in the care of his servant, Richard Lander, who returned to the coast and Fernando Po then went on to England in 1828.

RICHARD LEMON LANDER

Commissioned by the British government, Richard Lander and his brother John returned to Africa two years later to "claim" the glory of tracing the source of the Niger River to its culmination in the Atlantic Ocean. The Lander brothers landed at Badagry, just north of Lagos, in January 1830. They proceeded across land until they encountered the Niger near the city of Bussa, where Park had been killed thirty years earlier. They succeeded in securing a canoe at Bussa and they traveled with the river's current for weeks. After bouts of sickness and encountering innumerable hardships, they reached the Niger River delta in November 1830.

At this point the brothers were attacked by the indigenous people. Their canoe was plundered, then upset, throwing the brothers into the river. They were pulled from the water, stripped naked and put into chains. They were now not only prisoners but slaves. Finally, a minor chieftain called King Boy told them he would transport them to an English brig lying off the mouth of the river, provided the captain of the vessel would pay for their freedom. The captain initially agreed; however, when the brothers were on board the vessel, he refused to pay and forced King Boy off the ship. They then sailed to Fernando Po, where Consul Beecroft arranged for their passage on a ship bound for Rio de Janeiro. There, they were given space on a transport bound for home. They arrived in England on June 9, 1831.

The brothers could lay claim to two distinctions among the African Association's explorers. Despite having made the last part of their journey in chains they were the first to verify the termination of the Niger, one of Africa's great rivers. And, unlike most of their contemporaries, they both lived to tell the tale. In 1831 the African Association had been absorbed by the newly formed Geographical Society of London. In 1832 Richard Lander became the first winner of the Society's gold medal for his "important services in determining the course and termination of the Niger."

With the successes of Park, Clapperton, and Lander a path had been established that would allow Baikie to succeed twenty years later. Park had been the first European to view the Niger and had traveled far enough to determine that it flowed into the Atlantic. Clapperton laid the groundwork in the Sokoto caliphate that would allow Baikie to establish his successful trading ventures and Lander had traveled to the delta and established the exit point on the Atlantic Ocean of this great river. Inspired by the rising value of African markets and raw materials, British merchants now had first-hand knowledge of the Niger River and could exploit its full economic potential. Although these early explorers were driven by a quest for scientific knowledge, the value of their expeditions had been to gather useful economic knowledge of the interior. With this economic edge, British commercial interests might now be in position to bypass the powerful coastal societies that had monopolized Africa's international trade for centuries.[75]

Baikie would initially see his mission as charting unknown territory and collecting specimens of hitherto unknown plants and animals. Although the expedition sponsor would be the Geographical Society of London, the financial backer would be Macgregor Laird. As one of the greatest merchants of his time Laird had only one motive in supporting this venture, and that motive was profit.

CHAPTER FOUR

SETTING THE STAGE

For the first half of the nineteenth century, there was very little formal extension of British power in West Africa. Merchants were encouraged to establish trading sites and Britain seemed to be moving towards a laissez-faire policy towards trade with Africa. The cost of extending British rule seemed disproportionately great in relation to potential commercial success and the government began to abandon their early acquisitions, relegating their direct influence to a few strategic islands like Fernando Po.[76]

The only attempt at mapping the Niger River had been supported by the Saturday's Club. Now, the Geographical Society of London would send Baikie and others to refine the maps and establish trade within these recently charted areas. Most importantly, the major support for this effort would be led by an already established old African hand.

JOHN BEECROFT

During the second half of the 1840s John Beecroft was employed by the British government on various political and diplomatic missions in West Africa, and in 1849 he had been appointed consul for the Bights of Benin and Biafra. His primary responsibility was to keep an eye on slaving activities within the ports and maintain order among the often-warring factions within the Oil Rivers region near the delta. Beecroft was a very forceful man and Dike writes, "it was he who laid the foundations of British power in Nigeria and initiated the politics which were to characterize the consular period of Nigerian history."[77] His very remarkable career in West Africa had begun in 1827 on Fernando Po, when it became the base of operations for the suppression of the slave trade. When the island reverted to Spanish control in 1843, the slave squadron left, but Beecroft stayed and was appointed governor of the island by the King of Spain. In this latter posting he had become extremely powerful and influential and had won almost universal respect among the leaders of the coastal people.[78] On hearing about the proposed expedition, Beecroft offered to lead the mission. As a veteran of years of African exploration, he had no peer and was clearly the best person to command the venture.

The Secretary of State for Foreign Affairs accepted Beecroft's offer and immediately granted him permission to be absent from his consulate. Following the mission, he would return to Fernando Po and resume his consular duties. Baikie would return to England, bringing with him the journals and collections connected to the expedition and report to the Admiralty upon his arrival. Baikie was pleased. This would be his first voyage to Africa, and with Royal Naval personnel in charge of the ship and a noted explorer in the leadership role, he should have little to do but capture his specimens or collect and catalogue them.

HEINRICH BARTH

In addition to charting, mapping, selecting trading sites, and collecting specimens the expedition was directed to try to locate Dr. Heinrich Barth. Barth's expedition was funded by the British government, but the Geographical Society of London was a sponsor of that expedition as well. The famous German explorer had been traveling in Africa since 1849. Barth had been given a specific charge by the British Foreign Office to create maps and charts from Tangier in North Africa to Kano in West Africa. He was also to attempt to contact the Fulani leader in Sokoto.

Born in Hamburg in 1821, Barth had read classics at the University of Berlin. He was a scholar who was interested in the history and culture of the African people, rather than the possibilities of the commercial exploration of Africa. He was to become known as one of Africa's greatest explorers, Nigeria's foremost historian, and the man who constructed the frame of reference by which all later historical work would be written and judged.[79] The Society had agreed to support any publications that came from his venture and they, along with the British government, were interested in verifying his location and providing him support, if needed.

On his part, Barth, along with Englishman James Richardson and fellow German Adolf Overweg had landed in Tripoli and by 1851 had proceeded south as far as Lake Chad. Barth and Richardson were at odds from the beginning of the journey and the two decided to part company. From this point forward, Overweg and Barth sometimes traveled together and sometimes set out to explore on their own. Following one of their separate journeys, the two reconnected at Kukawa where Barth found Overweg "seized with a terrible fit of delirium and muttering unintelligible words." Overweg died the next day at the age of twenty-nine.[80] Subsequent reports were received by the Foreign Office stating that Richardson also had died. Barth was now on his own.

The last report on Barth, received in the early autumn of 1853, said he had entered Timbuktu. The report also indicated he was in the process of leaving and was planning to head overland to get to the coast. The proposed expedition in which Baikie would serve was to try to intercept Barth as he traveled along the river and to bring him to Fernando Po or bring him back with them to England if he so desired. Baikie readily agreed to participate in the mission. It is not known if he was aware that this would be the third attempt to explore the Niger River by traveling upstream since the termination point had been discovered by the Lander brothers. Nor is it known whether he realized what disastrous outcomes the first two attempts, by Macgregor Laird in 1832 and the Model Farm Project in 1841, had produced.

MACGREGOR LAIRD

Not surprisingly, the crucial discovery of the Lander brothers was hailed with more enthusiasm by the merchants than by the scientists and geographers. The long-sought highway into Central Africa had been found. The Niger could open the continent as the Rhine and Mississippi had done in their respective continents. Macgregor Laird saw a substantial commercial advantage for his company. In 1832 Laird was only twenty-three years old and exploring the Niger presented him with the irresistible charms of novelty, danger, and adventure. If it was successful, it would also provide an extensive market for his goods.

Profit and adventure were not the sole reasons for Laird's interest. He was also a staunch abolitionist. By introducing legitimate commerce into the center of the country, Laird believed they would strike a blow to the "debasing and demoralizing traffic which for years has cursed this unhappy land."[81] He convinced the British government in his belief that the steamship developed by his brother, John, was the key to opening Africa to Christianity and trade. He wrote to Lord Gray, "Steam powered ships will convert a most uncertain and precarious trade into a regular and steady one, diminish the risk of life, and free a large portion of capital at present engaged in it." His plan was approved, but his first journey into the Dark Continent met with disastrous results.

Laird left England aboard the *Quorra*, along with twenty-six British crewmen. They were accompanied by the *Alburkha,* with its complement of fourteen additional countrymen. No one knew the Niger better than the Lander brothers, and Richard Lander had been convinced to join the expedition. This decision would cost him his life. The British crew had been supplemented by hiring men from the indigenous Kru people along the coast. This group of workers were from the coastal area of what would become Liberia and the Ivory Coast, and they had worked for wages on European ships long before the abolition of the slave trade.

Among the coastal people, the most well-known sailors were the Kru and many them would be employed for Baikie's voyage as well. They were excellent boatmen and sailors and their dexterity on the water was quite astonishing. Hiring Kru sailors was a common practice with all vessels as they had a reputation of being both hardworking and reliable, as well as the best sailors along the coast. More than one British captain had gone on record saying that he considered the Kru superior to white sailors. But even these experienced sailors had not been able to save Laird and the others from Africa.

The sailing master of the *Quorra* was Captain Harris. He was the first to be laid low from fever and Laird had been forced to assume his duties. Dr. Briggs, the ship's surgeon, was another early casualty and junior surgeon R. A. K. Oldfield had been forced to assume all Dr Briggs's medical obligations. The winding course of the Niger delta swamps had made progress slow and most of the deaths occurred before the ships could even pass through into the river itself.

The voyage had taken them to the confluence of the Niger with the Benue, but with dire consequences. By their return there had been twenty-four deaths on board the *Quorra*. Of the Englishmen, Laird was one of only five survivors. On the *Alburkha* only Oldfield and one British seaman lived long enough to return to the Atlantic. In total, thirty-eight of the forty-eight Englishmen had died from tropical diseases. Richard Lander, and a few sailors, had been scouting ahead in a small boat when they had been attacked by local warriors. Lander had been wounded. He had developed an infection and fever and had died before the medical officer could tend to his wounds.

Laird become extremely ill, returning to England more dead than alive. He suffered from recurring bouts of fever for the next twenty years, never fully recovering from the effects of the expedition. Oldfield, whom Baikie was scheduled to meet later during his voyage, had also survived and now lived permanently in Sierra Leone, having abandoned medicine to make his fortune in trade. Although Laird had never returned to the continent itself, he never lost his interest in Africa, or his love of and respect for Africans, and his firm conviction that the development of legitimate trade was the only realistic way to end slavery.[82]

THE MODEL FARM

The loss of almost all the British seamen on Laird's voyage had discouraged many others from attempting additional surveys. Nevertheless, a second disastrous voyage on the Niger, called the Model Farm Project, was planned in 1841. This expedition was intended to implement the agreement that Clapperton had negotiated in Sokoto in 1824. The confluence of the Niger and Benue was the location that the Omani Muslims thought was the start of the Atlantic Ocean because of the extensive floods that occurred there during the rainy season. Clapperton had called them the Lakes of Nupe and had accepted what he was told—that they were connected to the Atlantic.[83] Thus, the confluence of these great rivers was for the second time planned to be the focal point for West African settlement and to serve as a center for trade by Great Britain. At this time the Model Farm Project, promoted under the auspices of Prince Albert was introduced to the public. It began with Albert's meeting with the "Society for the Extinction of the Slave Trade, and for the Civilization of Africa" on the June 2, 1840. The following is taken from a summary of that meeting:

> First Anniversary Meeting of the Society for the Extinction of the Slave Trade, and for the Civilization of Africa
>
> The public announcement that his Royal Highness Prince Albert, who a few days ago accepted the office of President of this Society, would take the chair at its first meeting, caused the demand for tickets to be unparalleled.
>
> Dr. Lushington, MP, then came forward to move the resolution— "that the utter failure of every attempt by treaty, by remonstrance, and by naval armaments, to arrest the progress of the trade, and the exposure of the deep interest which the African chiefs have in its continuance, as the means of obtaining European goods and manufactures, prove the necessity of resorting to a preventive policy founded on different and higher principles." (Loudly cheered)

> The Venerable Archdeacon Wilberforce concluded by moving the second resolution. The resolution was put to the meeting and passed unanimously.

In front of its largest gathering ever and under the guidance of the Prince Consort, the Anti-Slavery Society declared that the only complete cure of all the evils that the slave trade had caused was the introduction of Christianity into Africa. This Model Farm would be directed by Englishmen and supported by freed slaves and would afford essential assistance to the local people by furnishing them with useful information as to the best mode of cultivation, and the best crops to plant to ensure a steady market by introducing the most advanced pieces of agricultural implements and the hardiest seeds. Thus, the Model Farm would provide goods that could be traded directly with the indigenous people, freeing them from their reliance on slavery as a source of revenue.

The first task was to build three iron bottom ships that would be used for the mission. By the end of August in 1840, two of the three iron bottoms had been launched. One was christened the *Albert.* The name of the second vessel was the *Wilberforce*, as a tribute to the famous anti-slavery campaigner. The final ship was called the *Soudan,* and it was launched in the early fall of 1841(Figure 4.1, *see colour section*). Command of the expedition was given to Captain H. D. Trotter of the Royal Navy. He would be assisted by Captain William Allen, also of the Royal Navy, who had been with Laird in the ill-fated voyage of 1832.

When the ships had been built, a general announcement was made in the shipyards, offering an opportunity to serve on the project. The twenty-two English crew members were recruited from those who had built the ships and who were now looking for work and possibly adventure. Because of their construction and building skills, it was thought they would be instrumental for building the houses and outbuildings at the site of the Model Farm.

Not everyone, however, was in favor of this venture and the most vocal of the critics was Macgregor Laird. A farewell meeting was held in Exeter Hall and Laird tried one last time to voice his concerns. In front of a packed audience he said that his overriding fear was that the failure of this mission and loss of life on a grand scale would "so upset the British public that the very name of the Niger and all subsequent plans for exploration and development would be stopped for generations." After he had passionately stated what he believed had to be said, he was hooted and howled down. Laird was demoralized by this lack of respect. He had been the only one in the hall who had been on the Niger River. He had witnessed firsthand the terrible toll that disease could and would take. Yet in attempting to convince those gathered in Exeter Hall, he was fighting a lost cause.[84]

By the time the three expedition ships arrived at the delta, the fever had already begun taking its toll. Yet the determined party pressed on and negotiated a treaty with the ruler of Idda to purchase a site that for the Model Farm. The land agreed upon was located at the confluence of the Niger and Benue rivers and payment was made in cowry shells, the currency of the region. The English merchants thought they had won a bargain, as before they had set sail they had dredged the shells by the ton and packed them into the hold of the ship. However, land in West Africa at that time was considered communal property and was not within anyone's power to sell. So, the ruler was paid for selling land he did not actually own.[85]

Packet ships returning from Africa brought news updates from the expedition. The following excerpts from the *London Times* paint a bleak picture:

> *The Times*, November 11, 1841
>
> Captain Martin, of the *Daedalus* arrived at Liverpool yesterday from the coast of Africa. Captain Martin reports that the Niger expedition entered the new branch of the Niger between the 13th and the 15th of August, the *Soudan* leading. The expedition had, up to that date, lost nine hands by death.
>
> *The Times*, November 21, 1842—The Niger Expedition
>
> Extract from a letter dated Cape Coast Castle, September 26, 1842
>
> The *Wilberforce*, you will recollect, was here in March last, at which time Captain Allen was preparing to re-ascend the Niger, to look after the "Model Farm" people, and if possible to do something to retrieve the fame of the expedition. He proceeded hence to Fernando Po, to fit out the *Soudan*, to accompany him. While he was still lying there the *Kite* steamer arrived with orders from Government that only one vessel was to go up the river, and that she was only to have on board four or five white men at most. Her only object in going up was to be the bringing back the people left at the farm. The *Wilberforce*, under charge of her present commander (Lieutenant Webb), proceeded up the river, and found the "'Model Farm" a very perfect model of disorganization. The blacks who had been left at it, having plenty of cowries (a species of India shell used as money) and goods, voted themselves to be independent country gentlemen, and managed to get hold of a lot of natives, whom they very coolly made slaves of, and whom they compelled to work on the farm, each gentleman being provided with a *cat*, or slave driver's whip, the better to enforce obedience. The model farmer himself (Carr, brother of the Chief Justice of Sierra Leone) has never been heard of, and had, as it afterwards appeared, been killed somewhere near the mouth of the river. The *Wilberforce* brought away farm implements, people and all, and those of the latter belonging to this place are now being discharged here. The steamer got on a rock in the river, where she remained five days, and came down with a hole in her bottom, which now compels her to go home. So much for the last speech and dying words of the far-famed Niger expedition. A more mismanaged piece of business from beginning to end is not, I will venture to say, to be found recorded in any history.

At least two of the supervisors appear to have been killed by African workers who had been brought in to work the farm. Frightened, panicked, and to avoid being killed, some of the Europeans locked themselves inside their quarters. Others hid in the brush for several days until the *Wilberforce* returned to pick them up. Most of the survivors returned to England in poor health and with no desire ever to return to Africa.

Sending this massive expedition up the Niger had cost the British government £100,000 (approximately £10,000,000 in today's money). They had sent 145 Europeans and 133 African auxiliaries. These included interpreters, stokers, laborers, an African chaplain named Samuel Crowther, and Simon Jonas, his Igbo interpreter. Within two months the expedition had lost forty-eight European members and dozens of African workers.[86]

All of this might not have happened had Captain Trotter realized that he carried with him the medication that would make life possible for Europeans in West Africa. His ships had ample supplies of quinine, but Baikie's proof that the drug could be a preventative, and not simply a cure for those who had contracted the disease, still lay over a decade in the future. Trotter and Allen did not understand or appreciate the value of quinine, and consequently directed that it be administered only when the patient was already showing signs of recovery.[87]

By most accounts, the 1841 expedition would be correctly viewed as a complete failure. The loss of lives and the expenditure of vast amounts of money did not provide even a short-term gain. However, the presence of one man on this voyage would dramatically alter missionary work in Africa. Samuel Ajayi Crowther, a Yoruba ex-slave and recently ordained a minister following his studies in England, was a member of the exploring party. Simon Jonas was Crowther's interpreter, and through Crowther's preaching "heard the calling." Jonas would become known as the "first Igbo apostle" and served as a missionary for the remainder of his life. Their successful participation on behalf of the Church Missionary Society (CMS) enabled this organization to make a radical change in the Society's approach to Christianizing Africa. From this point forward, educated Africans; and not Europeans, would be used for those evangelical activities in West Africa. And this same Samuel Crowther would become a major player on both of Baikie's subsequent voyages.[88]

THE AFRICAN STEAMSHIP COMPANY

Laird had first looked to America as the location to establish the British and American Steam Navigation Company. This firm operated between Britain and America between 1839 and 1841. In 1841 the *President,* one of four ships built for the transatlantic run, sank with the loss of all those on board. The same year Samuel Cunard, the Halifax shipping magnate who founded the Cunard Line, had successfully secured the mail contract between England and America. These two factors had forced Laird into bankruptcy. Cunard was looking to Africa but Laid would not be defeated a second time. He would bounce back by pouring all his labor and capital into his African venture.

The Royal Charter, granted to the African Steamship Company in 1852, held the directors responsible for forming a company to establish and maintain postal communication between Great Britain and the West Coast of Africa. The charter further stated that the conveyance of the mail would involve a monthly service to the islands of Madeira and Tenerife. Stops along the West Coast of Africa would include Gorée, Bathurst, Sierra Leone, Liberia, Cape Coast Castle, Accra, Whydah, Badagry, Lagos, Old Calabar, the Cameroons, and Fernando Po.

Five iron screw steamships would be built in 1852 and 1853 for this endeavor by John Laird, Macgregor Laird's brother. Macgregor Laird had spent the twenty years since the attempt to establish the Model Farm in 1841 building his reputation and fortune through his trade with Africa. He was a man of strong opinions, and it was because of his views on African development that Laird and Thomas Stirling had formed the African Steamship Company. Laird believed strongly that the purpose of trade in Africa was to replace the traffic in human flesh with the sale of goods like palm oil and cotton and that the means of doing so lay in the steamship. Sailing ships could not navigate the rivers. However, shallow draft ships, utilizing steam power, could allow travel into the interior to conduct legitimate trade where the collection of slaves was still taking place. Laird believed replacing human trafficking with a robust trade in palm oil or cotton was the first step in eradicating slavery. He was a staunch supporter of the British colonial system and thought that civilization could never be achieved in Africa by the efforts of the indigenous population alone. Laird believed to civilize Africa a mighty power like Britain, which could take the various African territories under her protection, was needed to help. Through the labor and industry of its population and the cultivation of its soil the people would gain wealth. If a British-protected area developed an economy that could sustain itself through agriculture and legitimate trade there would be no further need for British "protection".[89]

Laird and Stirling had calculated that the new wealth of Africa would create new profits for Britain and replace the revenue produced by slavery for Africans by exporting palm oil. This natural product could be made into soap, but, more importantly, it could also be used to lubricate the steam engines that were being manufactured all over industrial Britain. Laird believed that once this new economy had been established, Christian religion and organized government would follow. Only at that point could these emerging African countries be considered as independent trading partners with Britain. If Baikie's voyage proved successful, Laird's and Stirling's initial stake in the African Steamship Company would make them very wealthy.

It was Laird's money and his ship that would make Baikie's first voyage into Africa possible and he wanted to interview each recommended member personally. For this reason, Baikie traveled to Liverpool for his initial meeting with Laird. As Laird's voyage and the Model Farm Project had been widely reported, Baikie may have been aware of the two catastrophic previous attempts to explore and settle the Niger River area, although at the time of the initial voyage he would have been only seven years old, and during the time of the Model Farm Project he would have been beginning his university studies. It is probable that Laird talked, at least in general terms, about the hazards of African travel at their preliminary meeting. It is more probable that Laird discussed his personal experiences on the river in detail. The extent to which these stories affected Baikie is unknown and at this point, irrelevant. What is clear is that nothing that he heard dissuaded Baikie from continuing his mission, although at the time the role he was expecting to play was very different from the one he ultimately did.

CHAPTER FIVE

THE VOYAGE OUT

On May 24, 1854 Baikie was aboard the 321-ton packet ship, the *Forerunner* on the first portion of his journey. Packet ships were small boats designed for mail, passenger, and freight transportation between England and her colonies. The *Forerunner* had been designed and built by John Laird at his shipbuilding firm in Birkenhead and it was launched in 1852 for regular voyages between England and the west coast of Africa as part of the contract between MacGregor Laird and Thomas Stirling and the African Steamship Company. This was its seventh voyage since being launched, now under the command of Captain Thomas Johnstone.[90]

William Balfour Baikie already looked much older than his twenty-nine years. He was unusually tall and had a head of thick, dark curly hair and a full beard (Figure 5.1). His Royal Navy uniform was perfectly tailored. He was described by his friends and family as a deliberate and thoughtful man. Baikie had been assigned to serve in a variety of roles in the expedition but on this portion of the voyage he was simply another passenger.

JOHN DALTON AND WILHELM BLEEK

As Baikie tracked the loading of the cargo he counted eleven men and one woman among the passengers who would be traveling with him to Africa. He noted with special interest two of them, Dr. Wilhelm Bleek, the etymologist, and Mr. John Dalton, the zoological assistant. These were the only two members of the expedition who would be traveling with Baikie on the *Forerunner*. Dalton was assigned to work with Baikie, collecting and cataloging specimens once they had reached Africa. The position of Bleek in the chain of command was less clear. He was twenty-seven years

Figure 5.1 Baikie at the time of his first voyage © Orkney Library and Archive

old and he seemed to Baikie to be suffering from either fatigue or pain. Bleek was stocky and his long brown hair fell almost to his shoulders. He had a moustache and full beard that almost covered his entire face. In contrast, Dalton was clean-shaven with close-cropped hair and had just turned twenty-one.

The three had met only two days earlier. Baikie knew John Dalton from his academic record at the University of Edinburgh. He had spent the last several weeks acquainting himself with Dalton's work and found him very capable. However, Wilhelm Heinrich Immanuel Bleek had been a last-minute replacement for the senior surgeon, Dr. J. W. D. Brown, who had been transferred to the Royal Fleet that was already heading toward the Crimean War in the Baltic. All Baikie knew of Bleek was contained in the thin file he had been given at their initial meeting. Bleek had graduated from the University of Berlin three years earlier with a doctorate in linguistics. When he was in Berlin he had become interested in African languages, which became his thesis topic. Although he was a student of Africa, this would be his first trip to the continent.

The linguist Bleek had replaced the surgeon Brown, and hence his role was unclear. Dr. Baikie had been appointed as assistant surgeon and naturalist on the proposed expedition. This would have given Baikie most of his time free to pursue his role as naturalist. Now, with Brown shipped off to the Crimean War, it appeared that Baikie would be the only medical officer accompanying the expedition. Bleek was a civilian who had no training in medicine or in natural science. This must have raised questions in Baikie's mind over what Bleek could offer to the goals of the mission, and how this change would affect both Baikie's role and the chain of command.

After the crewmen cast off the lines the *Forerunner* moved quickly away from the dock. The ship was a fast screw vessel, allowing it to be propelled by its steam engine or to work under sail. Under the full force of the powerful engines they were quickly transported into open water.

Before the ship had left port, Baikie had discovered that one of his fellow passengers was a Mr. Louis Fraser, a survivor of the Model Farm Project in 1841.[91] Fraser was an experienced naturalist who had worked for fourteen years at the museum of the Zoological Society of London. At the time of the Model Farm expedition he was working with the American Civilization Society, a group promoting emigration of American ex-slaves to Liberia. Started by American Speaker of the House Henry Clay and Supreme Court Justice Bushrod Washington, this prestigious group also included future President Andrew Jackson. Intended to solve the problem of America's growing number of freed Africans, the project was unsuccessful and fewer than 3000 African Americans ever left for Liberia. Following the disaster of the Model Farm, Fraser had returned to his work as a naturalist.

While the other European survivors elected to return to England as quickly as possible, Fraser stayed on in Fernando Po and collected specimens for various English museums and gardens. Baikie was familiar with *Zoologica Typica,* a lavishly illustrated book that Fraser had published in 1849. In 1850 Fraser had been appointed consul at Ouidah (located in present day Benin), a position he had vacated shortly before the *Forerunner* left Plymouth.[92] Few white men knew the Niger River basin as well as Fraser and Baikie was determined to gather as much information as possible about the river and what he might encounter from his traveling companion during the voyage.

That night, after the two had been speaking for hours, Fraser retired. But Baikie remained on deck until it was nearly dark, and, except for the helmsman and forward watch, the deck was deserted. The ship was now under full sail, its engine quiet. He was on the brink of the most significant assignment of his career and watching the bow of the *Forerunner* slicing toward Africa fostered many thoughts. He had eagerly taken on this assignment and had even lobbied to be accepted. He had stepped onto this ship in eager anticipation of finding new species of plants and animals, and of sharing those finds with his museum colleagues at home. A young man who is not yet thirty years of age seldom considers his own mortality. But for Baikie, having heard Laird's stories of the earlier failed missions and after his recent conversations with Fraser regarding the massive human toll associated with the Model Farm Project, the the dangers of Africa must suddenly have seemed very real.

The following morning, Baikie rose early. He climbed onto the deck and walked to the stern of the ship. In the early morning light, the sea seemed unusually calm and only the sounds of the engine broke the silence. Although his official duties would not begin until he was on the Niger, this seemed like an ideal time to test some of the equipment and to collect samples of marine life that would surely be following the ship. Baikie returned to the hold, retrieved his towing nets, and brought them back up on deck. He checked the securing line, making a coil at his feet. He then wrapped the rope around his wrist and forearm several times. Almost as an afterthought, he secured the remaining line to the rail of the ship.

His cast took the net far astern and for a few seconds the line played out slowly. Suddenly the line shot forward. The coil at his feet instantly disappeared with a hissing sound and the section of line wrapped around his arm took the full force. He straightened his arm so that the line could be pulled free. The last of the rope flew from his arm, burning a stripe across the sleeve of his uniform coat, and only the end tied to the rail prevented him from losing all his gear.

He sat down on the deck; his right arm draped gingerly across his lap. The calmness of the surface had fooled him. He had misjudged the force created by the nine-knot speed of the ship. His job on this expedition was to collect specimens and this first attempt had almost resulted in the loss of all his equipment. He slowly retrieved his nets, doing most of the work with his left arm. The young, highly educated Dr. Baikie would learn a great deal on this voyage.

Shortly after dark, on the fourth night out, the engines ground to a halt. While repairs were being made, the *Forerunner* proceeded under sail. The pace had slowed to two knots and Baikie decided again to try his luck with the drift nets. Securing the nets and lines from the hold, Baikie coiled the rope at his feet and tied the end to the rail.

With the slow speed and light winds, the net worked perfectly and each time it was retrieved it yielded various snails and jellyfish. Each cast produced a treasure and Baikie remained at this task for most of the night. The jellyfish were phosphorescent and especially mesmerizing. Out of curiosity, Baikie held several near his pocket watch and found he could easily read the dial. A few of the specimens were bottled, labeled and recorded in one of the large journals purchased for this practice. The rest were thrown back over the side to follow in the wake, no worse for wear. When Baikie finally grew tired of collecting he sat on the deck propped against the center mast as schools of porpoises played around the ship. Their presence indicated that the *Forerunner* was nearing warmer waters and that Baikie would soon be in Africa.

After six days of continued calm seas, the ship came into sight of the island of Porto Santo. There was no intention to land, as the *Forerunner* was soon to be in Madeira. However, this being the first land glimpsed in six days the sighting created a great deal of interest among the passengers and most were soon gathered along the rail of the ship.

Along the north side of the island high limestone cliffs descended almost to the water line, separated from the ocean by only a narrow stretch of beach. Down the eastern side of the island the cliffs fell away and were replaced by flat lands, sparsely covered with brome grass. Farther inland cattle were grazing in the fields and rows of a vineyard could be seen. As the ship moved toward the south end of the island, the vegetation was replaced by a magnificent beach of golden sand. Finally, the *Forerunner* rounded the point at the end of Santo Porto, turned, and headed for Madeira.

CHOLERA QUARANTINE

Baikie knew that in this part of the Atlantic the various ports had been settled at a time when the Spanish and Portuguese sailors were working their stepping-stone approach to discovering the route around the tip of Africa. Their method was to travel until they found an island with a suitable harbor and would then create a shipping station there. The next captain would travel to that base and then extend only to the next useable port. This was at a time when European sea captains were still concerned about sea dragons and falling off the edge of the earth. These fears meant that the island ports were close together and Madeira was soon reached by ship.

The *Forerunner* was soon at anchor in the harbor of Funchal. No sooner had the ship dropped anchor than the passengers once again appeared on deck. In ones and twos, they ventured out into the sun with anticipation. They were formally dressed in hats and coats and clearly had been eagerly preparing for their excursion on shore. Most of the passengers had never been on such a lengthy sea voyage, nor had they even been out of sight of land. Even Baikie, with his extensive experience in the Mediterranean, felt a sudden urgent need to step onto dry land.

Baikie knew from his previous voyages with the Royal Navy that each time a ship made a port of call the local health authorities were required to sign a permit allowing the vessel to remain at anchor and to discharge passengers and cargo. This was generally a formality and often an excuse for the local port authority to collect a fee without having to provide a service. However, this time it proved to be more than just a formality.

When the health boat appeared alongside, the crew and passengers were informed that an epidemic of cholera had broken out in Glasgow. Even though the ship had been nowhere near Scotland, the passengers would not be allowed to leave the ship. Some shouted at the *Forerunner's* captain, trying to make him intercede. Others begged the Portuguese health officials to let them go ashore. The efforts were of no avail.

Other action in the harbor; however, was heating up and the *Forerunner* was soon surrounded by various vessels. Coal barges came alongside and began discharging their cargo. Smaller boats with articles for sale vied with each other to get as close as possible to the side of the ship. Tradesmen hawked their wares to the limited audience on board. Bunches of cherries were bought at sixpence each. Other venders sold baskets, mats, and straw hats of local manufacture, with every item going for a shilling apiece.

In making the acquisitions while in quarantine, buyers could examine the goods only by sight. If they touched the item, like it or not, they would have to keep it. Baikie noted with some amusement that the sellers had much less concern about handling the money used for the sale. When a price was agreed upon, the seller would hold up a small cup containing water into which the customer dropped the money. Without hesitation, the seller would dump the water and pocket the money. Baikie noted, "While individuals and property of all description had to undergo detention, purification or fumigation; the contagious properties of the coin must have been immediately destroyed by simply passing it through sea water." This routine continued until nearly sunset, and shortly thereafter the *Forerunner* set off for Tenerife. The passengers, including Baikie, stood by the rail until Madeira disappeared below the horizon. The long voyage suddenly seemed much longer.[93]

The *Forerunner* was sailing among the first of the Canary Islands by early morning and Tenerife was sighted the next afternoon. At about eleven pm that evening the *Forerunner* was anchored off Santa Cruz, its capital. Tenerife, in the center of the Canary Islands, belonged to Spain. Free from the Portuguese health restrictions, the passengers once again hoped that they would be allowed off the ship. Because the water in the harbor was extremely deep, the ship was anchored very close to shore. On board, a mare was being disembarked to the governor of the island. She was brought on deck, nervously prancing, perhaps anxious about the next move, then, with little fanfare, she was urged over the side and made to swim ashore. The horse's harness was dropped into the arms of a Spanish sailor who had brought a rowboat next to the ship. Baikie believed that the unloading of the horse and the transfer of the tack were all signs that they would finally be allowed to step onto dry land. However, when the health boat finally arrived in the early hours of the morning the *Forerunner* was once again slapped into quarantine without further discussion.

INTO AFRICA

After leaving Tenerife, the ship ran towards the coast of Africa, with the northeast trade winds carrying the vessel at a steady pace under clear skies. Shoals of flying fish surrounded the boat, amusing the passengers with their glittering flights. Rising into the wind in various numbers from a half dozen to fifty or sixty at a time, and setting their pectoral fins like ridged blades, they would shoot from the ocean and remain aloft for from eighty to one hundred yards. Many flew onboard, where the passengers tried to scoop them up before they could wriggle back into the sea. Those that were captured were quickly handed over to the cook, who stood waiting with a large wicker basket. Even the young woman passenger was participating in the fun. While all her concentration was focused on capturing dinner from an earlier landing, one particularly good flyer struck her with such force that it nearly knocked her to the deck. However, she appeared more embarrassed than injured.[94] Standing on deck, Baikie realized that the air, which had hitherto been close and repressive, had become much cooler. In the distance, large flocks of petrels could be seen approaching and following the ship. Baikie knew that the change in weather and the sighting of these dark-colored sea birds showed that Africa was close and the *Forerunner* was soon anchored off Gorée Island (now part of Senegal).

For the first time during the journey the passengers were allowed off the ship, with strict orders to be back on board within two hours. As Baikie stepped on to the shore the wind that had been blowing in from the sea suddenly shifted. The breeze now came from deep within the island and passed across the small village near the harbor. It was saturated with the scent of sweet flowers and decaying garbage. The smells of animals and people mixed with the aroma of mildew and rotting vegetation. It carried with it a wet smell that was both acrid and sweet and felt so heavy that it seemed he would have to pull it into his lungs. It was unlike anything Baikie had experienced before and it framed his first impression of Africa.

The *Forerunner* remained at anchor in the harbor at Gorée overnight and on the following morning moved to the mouth of the River Gambia. There the ship was struck by the first tornado of the season and it had to turn quickly into the wind and drop anchor. Although the storm had caused the sky to darken to an ominous green tint, it had brought only a slight wind and little rain. The weather had caused only a minor delay in the voyage, but more importantly, it marked the beginning of the rainy season.

Easy navigation on any river requires a wide, deep channel, smooth bends in its course and a low velocity of water. The Niger River has none of these. Its flow and current vary according to the uneven distribution of rainfall over its course. Differences of thirty-five feet between high and low water levels are common. Over most of its course the Niger has a two-season climate: a dry season from November to March, followed by five to seven months of rain. The length of time they could spend on the Niger would be dictated by the depth of the river, and the depth was based on the swollen currents provided by the seasonal rains. Baikie was aware of the time constraints the mission faced and the lost days that had already passed. A few simple computations of the time needed in preparation before they could begin their voyage upstream told him that, if his calculations were correct, they were already well behind schedule.

The *Forerunner* was again underway and heading for Sierra Leone when the second tornado appeared. The ship was about thirty miles offshore and initially the storm could be seen traveling parallel to the coast, between the ship and the land. It looked as if it would not cross the ship's course but would move inward across the African coast. But suddenly the storm's direction altered. The huge cloud arched, showing a dark mass high in the heaven and producing a lighter grey-green sky underneath. It seemed to leap forward directly toward the ship, pushing a low white line of foam in the sea at its head. Just before the storm hit the *Forerunner,* the temperature dropped twenty degrees Fahrenheit and there was not a breath of air. Within an instant the storm burst over the ship and the fierce gale was accompanied by tremendous rain, the drops hitting like hail, burning the face and arms of anyone foolish enough to remain on deck. There was no thunder or lightning, but the fury of the storm lasted for over an hour. The violence of the rains beat the sea until it was flat. As the storm began to subside the sea again sprang to life. The waves swelled to over six feet and the ship's pace was slowed to a crawl. Baikie had been in intense storms while on the Mediterranean and had witnessed the North Sea winds howling through the Orkney Islands, but he had never witnessed a storm with this violence or intensity.

SIERRA LEONE

After several days sailing, the *Forerunner* made port at Sierra Leone at night and the next morning Baikie was up with the sun. Sierra Leone was one of the first British colonies in western Africa and at the end of the 1700s the British had prepared Freetown to become a refuge for freed slaves. However, like the Model Farm, the grand experiment had only proven how difficult it was to transplant people and ideas into cultures other than their own. This town had initially received several hundred "freedmen," who had been servants of the wealthy in Britain. These former slaves had been footmen, valets, and butlers who had lived most of their lives in British manor houses, experiencing the second-hand luxury that their stations provided.

Along with about sixty prostitutes from London and Portsmouth, the mixture of groups settled in Freetown proved toxic. It had been thought that the British-influenced Africans returning to their homeland would prove the ideal core population on which to rebuild Africa in a Christian image. It was anticipated that after settling in Freetown for a period of time, these transplants would return to their various former homelands throughout western Africa and both "civilization" and Christianity would be spread throughout the continent. However, disease and hostility from the local Africans nearly eliminated this first group of "immigrants."

However, shiploads of slaves rescued by the West Africa Squadron were also liberated and taken to Freetown. Against all odds the settlement had not only survived but had also prospered. The later groups of people had proven to be far more able to deal with the climate and their neighbors than the first ones. But it was the lack of their dispersal to other parts of Africa that had been the failure. Most of these transplants never left Sierra Leone. These returned Africans—or Creoles, as they came to be called—were from all areas of Africa. Cut off from their homes and traditions by their experience of slavery, it had proven much easier for them simply to assimilate to the British style of life installed by the original freedmen than to move to other regions. However, although they had failed to spread out and civilize Africa, their colonial sponsors were delighted that these individuals had developed a strong Christian tradition and built a flourishing trade center in Freetown.

Baikie could see that Freetown was built on sloping ground running from high on the hills to the river's edge. Closest to the river were large buildings that appeared to be used as storehouses. Freetown was a vital link in the expedition and Baikie had been given a specific task to complete while in port. He left the *Forerunner* and moved among the warehouses along the river. Traversing the wooden walkways proved extremely difficult as he continually had to dodge pedestrians and all manner of animals and vehicles. Goods marked with the names of British and European companies were bound for the warehouses. Newly arrived packages were trading places with bundles of elephant tusk and containers of groundnuts and palm oil that were heading toward the ships.

Baikie soon came to the establishment that was the goal of this quest. Above the door, a large sign in faded letters read "Oldfield's, Ltd." This was the same man who had traveled with Laird in 1831. It was the same Dr. Oldfield who had been one of only two European survivors onboard the *Alburkha,* and he had elected to remain in Africa as Laird's partner. As Baikie entered the door he was greeted by the smells of industry. The familiar aromas from kegs of rum and raw tobacco were mixed with the sickly-sweet odors of palm oil and human sweat. Commerce buzzed in the dim interior as a great number of Africans transferred goods from a large pile on the left side of the warehouse to an equally great mound on the right. Numbers of men were attempting to move the merchandise from the pile on the right through an oversized door at the rear of the warehouse. A third group was bringing commodities in through a side door and depositing them into the stack on the left, and thus the cycle continued.

At the front of the building two men were huddled over a ledger. The first introduced himself as Robert Oldfield and the second was presented as Ernest Heddlein. Oldfield had been asked by Laird to secure the interpreters who would be needed for the expedition and Baikie had been given the money to pay those whom Oldfield selected for their services. Oldfield had carefully chosen each of the men who would serve as crewmen on the vessel. They would also be asked to serve as interpreters. He provided Baikie the list of those he had found, and he seemed to have selected a group of well-qualified men for the journey. Baikie was assured that each of them spoke English and next to each name was listed that man's first language: Igbo, Yoruba, Hausa, Nupe, Bornuese, or Kanuri. The men would prove necessary as the expedition moved along the Niger and passed through the homeland of each of their people.

The primary transaction completed, Baikie welcomed the offer from Heddlein to walk along the shore and observe the commerce. He was enjoying being on land and knew that several more days at sea still lay ahead. As they moved along the side of the warehouse, Baikie could see that the two doors he had observed earlier both led to the river. At the water's edge a large canoe, over fifty feet in length, had been pulled as close to the shore as possible. Narrow wooden planks stretched from the shore and rested on the bow and stern of the vessel. Workers were taking goods from the front of the canoe and carrying them into the door at the side of the building Baikie had just left. Another group of men were coming from the second door and loading outgoing wares on to the rear of the same vessel.

This pattern, and the one he had observed earlier, completed the large circle of Laird's and Oldfield's activities in Africa. Ships coming from Laird's headquarters in England brought manufactured goods for sale in the Freetown shops and markets. At the center of the river they were offloaded from the seagoing ships onto these large cargo canoes. Having been transferred into Oldfield's warehouse, they were then sold to merchants and shopkeepers in the city. The profits from those sales were used to buy the raw materials of Africa. The outward-bound and inward-bound goods were stored in the same warehouse until the former could be loaded onto the canoes and transported to the now empty ships at the river's center. Palm oil collected in Africa would be taken back to Laird in England and sold to manufacturers, where it would grease the wheels of industry. The other raw products would be made into finished items that could be sold in the markets on the return trip to Africa. Oldfield, Heddlein, and others had developed a system that utilized a minimum amount of transportation yet maximized their profits on each leg of the voyage.

Baikie and Heddlein moved away from the river and its warehouses toward the retail area and the central market. Down every street an incessant stream of people flowed from one end of the city to the other. All along the route, shops were filled with eager purchasers and the central marketplace was thronged. People called out, some addressing Heddlein by name, urging the two of them to come and see their wares. Every ethnic group between Bornu and Timbuktu seemed to be represented in the market. This was borne out by the vast and colorful diversity of costumes. Many were dressed in the flowing tobes of the North Africans. These consisted of a length of cloth sewn into a long loose skirt; the cloth then fastened over one shoulder. Some wore ample turbans. Some were dressed in the European garments worn by the liberated Africans, and others in the seafaring garments of the Kru.

The market shoppers included European and Africans who could choose from a heady mixture of merchandise and African goods. However, each stall seemed to concentrate on selling only a single product. A booth selling only the very popular Dutch cloth with its blue background and white designs was next to a vender selling only carved ivory figures. Venders tugged at his coat, items for sale were thrust into his face, and the all-encompassing noise made conversation with Heddlein virtually impossible. Baikie was at once captivated and overpowered by the experience.

THE PLEIAD

Heddlein had said that they could spend as much time in the market as Baikie desired, but after forty-five minutes the climate, throngs of people, and the noise combined to exhaust him. As they were returning to the warehouse a small steamer could be seen entering the river. When it anchored near the shore, they saw it was the *Pleiad*. This was the ship that would carry Baikie's expedition up the Niger. Coming from Dublin, it had been following the route of the *Forerunner*. This flat-bottomed steamer had been built specifically to explore African rivers. Pleiad is a term that refers to a group of ancient Greek philosophers, and it was a strange name for a British ship, even for a noted eccentric like Macgregor Laird. It was a small screw steamer of 71 tons and was 105 feet in length. She was the first exploring vessel ever to be fitted with a new screw propeller and her 40-horsepower engine could reach a speed of ten knots per hour. With her propeller lifted, she became a fast-sailing schooner. Internally the ship contained five state rooms for the officers. The after cabin had mahogany tables, a leather sofa, a bronzed chandelier, and marble sideboards. The ship carried a large library, a steward's pantry, and a bath. The *Pleiad* was exceptionally well armed, with a 12-pounder pivot gun, four swivels, Minie rifles and double-barreled guns for the officers, and muskets for the crew.[95]

Baikie had explored every inch of the *Pleiad* in the harbor at Dublin but had not seen the vessel for over a month. Heddlein left him at the edge of the river after securing the services of a local sailor and a small boat, and Baikie directed the oarsman to the *Pleiad*. He boarded the vessel using the rope ladder dropped from her deck and came face-to-face with Thomas C.

Taylor. The *Pleiad* was under the command of a transfer captain, but Taylor would be piloting the vessel when it left on the expedition from Fernando Po. Taylor, like Bleek, had been a last-minute replacement. He was appointed when Lieutenant McLeod had also been called into service in the Crimean War. Baikie and Taylor had met briefly in Dublin where Baikie had been invited by Macgregor Laird to inspect the *Pleiad*.

Baikie informed Taylor that the *Forerunner* would leave as soon as he returned to the ship. Taylor responded that the *Pleiad* would leave the following day after laying in the last of its needed supplies. While the *Pleiad* needed to make only one more stop to obtain the rest of the Kru sailors required for the voyage, the *Forerunner* had to make numerous stops along the coast to discharge the other passengers and pick up outgoing mail and supplies. It was agreed that *Pleiad* should reach Fernando Po well ahead of Baikie's arrival on the *Forerunner*. Baikie was shuttled from the *Pleiad* to the *Forerunner* by the same boat that had brought him. It had been a brief visit, but Baikie's inspection reinforced his confidence in the vessel that was to be his home during the next phase of the expedition.

The next port of call was Monrovia. The winds were extremely high as they approached, and the seas were heavy. Up to this point, the *Forerunner* had been able to sail directly into the harbors or deep river ports. However, at the port of Monrovia and other ports along the Bight of Benin there were great deposits of sand stretching far out into the ocean and these sandbars make it impossible for deep-hulled ships to sail directly into the river channels. Canoes with a shallow enough draft to traverse the sandbars were routinely dispatched from land. These canoes would rendezvous with the ships outside the bars to onload or offload cargo and passengers amid the always-hazardous seas. In addition to the dangers of the canoe tipping in the open ocean and the passengers drowning in the waves, extremely large and aggressive sharks constantly patrolled the area of the sandbar looking, no doubt, for just such an unfortunate event.

Upon reaching the sandbar off Monrovia, Baikie observed a large canoe coming out from shore. It bobbed up and down between the waves. With the crossing of each crest the front of the canoe would crash back into the sea, a wall of spray flying to each side and then blown back over the front of the canoe. As the canoe dipped through the waves and moved closer, Baikie could see that the crewmen were all facing forward, six of them with paddles digging deeply into the rough surf. A lone passenger sat in the very front of the boat; and in this position was catching the brunt of the spray from the top of each wave.

The boat moved closer and was soon alongside the *Forerunner*. The rope ladder was lowered and one of the crewmen moved forward in the canoe to steady the passenger as he stood at the bow. Even from this distance it was evident that the ride from shore had resulted in him being soaked to the skin and, with the night coming on, he must have felt the bite of the cold breeze. A crewman waited for just the right moment, and when the canoe was at the top of a roller, he almost threw the passenger at the waiting ladder. Successfully making the grab, he was soon on board and was offered a blanket in which to wrap his shivering body. As the canoe pulled away from the *Forerunner* Baikie observed the two dark circling shapes next to the ship. They were no doubt disappointed that Jonathan Newman, the British Consul, was now safely on board. However, there would be other meals in their future.

Mr. Newman had been convalescing from an attack of remittent fever and his doctor had recommended a cruise around the Bight to improve his health. Baikie found it amusing that a doctor would recommend a treatment that would begin and end with a canoe trip that had the strong potential of killing the patient. However, securing his passage was the only reason for stopping at Monrovia and the *Forerunner* was soon again underway.

CAPE COAST CASTLE

On the morning of June 18, the ship arrived at Cape Coast Castle. Here the sea was unusually calm and the transfer from the ship to canoe, and canoe to shore was made without event. Baikie and the lady passenger were among the first to make the transfer. Upon reaching shore, she was met by a handsome young officer from the West Africa Squadron and at eleven o'clock that same day all the ship's passengers and crew were invited to attend her wedding in the fort's chapel. This was the first time that any of them had been aware of her reason for undertaking the voyage.

The Cape Coast was one of two locations on the voyage where Baikie had a previous connection. Charles Heddle was an acquaintance of his from his days in Kirkwall. He was also the nephew of Robert Heddle, Baikie's life-long friend and the co-author of "Historia Naturalis Orcadensis" that had been published during Baikie's first year in the Royal Navy. Charles Heddle now lived permanently in Sierra Leone and Baikie had planned to see him. However, with the delays in his schedule, the time was too short on outward bound portion of the trip to accommodate a visit. Baikie sent a note promising to visit Heddle and his family upon his return from the expedition. The Heddle family would play a continuing role in Baikie's life up until the time of his death.

On this visit to Cape Coast there was time to renew the second acquaintance. Baikie's mother had been Isabelle Hutton before marrying his father. John Hutton, Isabelle's cousin, was a very successful businessman who now resided in Cape Coast and that evening Baikie was invited to dine with him. Hutton lived near the ocean in a magnificent stone residence. Built high on a hill, overlooking the castle and the sea beyond, each of the large rooms had rows of windows nearly eight feet in height that were fitted with tight shutters to keep out the rains. On this evening the entire wall was opened to catch the soft night breeze.

During the dinner, they were waited on by five or six African children of varying ages. Baikie noted how reserved and well-mannered they were. Mr. Hutton replied that they had been a gift. He explained that after one of his dealings with the ruler of the Ashanti, he had been presented with a gift of seventeen children. This had caused him to face a tremendous decision. If he returned them, it would have been a personal affront to the King and would have effectively ended any future dealings. His refusal would also have ensured the children remained slaves and would simply have been sold or given to someone else. Therefore, Hutton had taken them with him. He first freed them. Then he clothed them, and fed them, and secured for each of them an education and his protection. As each of the children reached an appropriate age, they were given the choice of staying with him in his home or leaving to start their own lives. These six were those who remained of the initial seventeen.[96]

This was the first time that Baikie had been directly confronted with slavery. The thought that someone could simply "give" away seventeen children, as one would give a horse, was beyond his understanding. However, Hutton's acceptance of it, and decision to simply take them in, was also a solution he had not contemplated. How would he have reacted under similar circumstances? He would be traveling among cultural groups where this could possibly occur. Although he had yet to formulate what his response would be, he realized that he would need a plan to deal with such a situation sooner rather than later.

About sunset, just as Baikie was preparing to return to the ship, the winds begin to pick up. By the time he reached the beach a violent storm had set in and towering waves were breaking against the shore. It was well after eight o'clock before the storm had subsided enough to venture over the bar and he made his way to the beach where the canoes would be waiting.

Upon reaching the beach, Baikie discovered that the canoe men had waited out the storm by consuming large quantities of palm wine. Now, they were all either drunk or unwilling to set out. The *Forerunner* would be leaving within the hour and Baikie was concerned that he might be stranded. As the argument with the canoe men continued the master of an American vessel approached. As some of the canoe men were in his employ he asked if he could assist in any way.

The captain's forceful command that the canoe men row Baikie to his ship nevertheless made no impact on the men. At that point, the American pulled a revolver from the waistband of his trousers. Sobriety immediately returned to the boatmen and Baikie and was soon on his way back to the *Forerunner*. Half an hour later, Baikie was on board and the *Forerunner* was under steam and headed for Accra.

DEATH OF BEECROFT AND MEETING REVEREND CROWTHER

As at Monrovia, no one left the ship at the stop at Accra; they paused only long enough to take on mail. However, one of the messages taken onboard had the potential of ending the mission before it could begin. It brought word of the death of John Beecroft, Macgregor Laird's handpicked choice to lead the expedition. Baikie was at a loss. Would this mean the end of the voyage? Who among the party could possibly undertake to lead? Most of the members of the expedition were making their first visit to Africa and many were last-minute replacements for the original crew members. Due to the time of Beecroft's death, bringing someone in from outside was not feasible if they were to take advantage of the current rainy season; and no one at any point along the coast could match his credentials.

The voyage from Accra to Lagos was made amid constant heavy showers and occasional squalls. The rainy season had fully set in and the weather matched Baikie's mood. He was now contemplating leaving the ship at Lagos. With Beecroft's death the expedition would surely be scrapped or postponed until a suitable replacement could be located. Baikie was on leave from his Royal Navy assignment. His duties at Haslar Hospital were already being covered and with the cancellation of the expedition he would be free to spend the time exploring and collecting on his own schedule. He could possibly still make this voyage worthwhile.

Their arrival at Lagos was marked by lying at anchor for twenty-four hours. The recent storms and the continuous heavy swells created treacherous surf that rendered the bars of the river too dangerous to cross. Had the canoe put Baikie on shore upon his arrival he would probably have abandoned the assignment. As it was, the extra twenty-four hours to think allowed him time to firm up his original choice. He would continue to Fernando Po and wait for decisions that others would make.

With nothing to do but wait for the ocean to calm Baikie occupied himself by watching, Anna, the ship's dog, romp from deck to deck, attacking imaginary foes that only she could see. But not all of Anna's foes were imaginary. Since leaving Plymouth, Baikie had observed that Anna had been attempting to capture what seemed to be the ship's solitary rat. Day after day he would watch as the rodent would make its appearance to be chased at top speed by Anna, pursued from shelter to shelter, only to leap at the last minute into yet another hiding place. While they were at anchor, Baikie noticed that the rat had left its latest shelter and had ventured again on to the deck. This time Anna had the upper hand and appeared to have closed off all the usual avenues of retreat. The only option for the rat was to bolt into a scupper hole, which he did with such force that he popped completely through and dropped overboard. As Baikie watched, the rat swam bravely. But as the current carried the rat astern his struggles quickly ended with the appearance of a large shark. With one gulp the rat was swallowed, his disappointment no doubt exceeding that of Anna's.[97]

At last, the seas calmed enough for the canoes to be sent out from Lagos to the *Forerunner*. The arrival of the first canoe brought the Reverend Samuel Crowther. This was the same Reverend Crowther who had been attached to the Model Farm Project. Macgregor Laird had asked the CMS to allow Reverend Crowther to accompany the expedition, offering him free passage on the *Pleiad*. Crowther had been eager to go and the CMS was only too glad to agree.[98] Laird believed that commerce and Christianity should go hand in hand and that the trade centers would also make ideal bases for spreading the word of God along the Niger

Figure 6.1 Fernando Po © The Bodley Head and Random House

River. Laird found Crowther, with his extensive experience, an ideal addition to the mission and Reverend Crowther had traveled from Abeokuta to Lagos to join the expedition.

The Reverend Crowther was a short, solidly built man who appeared to be about forty-five years old. His skin was very dark and his round face was deeply creased from the tropical sun, but also from his constant smile. Baikie was a man who believed in first impressions, and Crowther appeared at first glance to be one who would be a welcome addition to the expedition. Despite his tattered cleric's garb, now drenched in seawater, his bearing was powerful and proud. He appeared to Baikie as someone who could be trusted.

From Lagos, the *Forerunner* continued down the coast staying within sight of shore. At this point along the coast, large rivers entered the ocean every few miles. Some of them were primary tributaries, others were branches of major rivers that had joined together many miles inland. The water along the shore was the color of mustard from the sediment deposited at the mouths of these various streams. Often the color extended several miles out to sea and the stench from the rivers' runoff was nearly overpowering.

By that evening, Baikie was standing on the deck of the *Forerunner*, at anchor in Clarence Cove on the island of Fernando Po. The high land of the island was backlit by the most beautiful sunset Baikie had seen since leaving Plymouth, but the newly made grave of Consul Beecroft at the edge of the bluff provided an ominous reminder of the problems ahead. A more immediate concern was the absence of the *Pleiad*. With the extra stops that the *Forerunner* had made along the route, the *Pleiad* should have arrived more than two days earlier. Baikie looked around the cove and could see three other ships at anchor. Unfortunately, none were the *Pleiad*.

CHAPTER SIX

THE COMMAND

I t had been two days since Baikie's arrival on Fernando Po and there was still no sign of the *Pleiad* (Figure 6.1). At the island's capital, Port Clarence, Beecroft's death was talked of everywhere. In the town, everyone Baikie spoke to was grieving over the loss of the man they respected as a kind-hearted friend and protector. James Lynslager had become acting consul upon Beecroft's death and he and Baikie were obliged to meet and discuss the fate of the mission. Moreover, upon reaching Fernando Po, Baikie had been informed that Heinrich Barth had been sighted on the upper section of the Niger River within the past month and, as he was now traveling alone, there was added concern for his safety as well. Based on the sighting of Barth, the decision had been made to continue the voyage despite the late start and Beecroft's death. Linking up with Barth somewhere near the confluence of the Niger and Benue rivers was now more important than ever. This would require great luck in timing but also someone with a solid knowledge of the lands surrounding the river.

The "Admiralty Instructions," hand carried by Baikie, were delivered to Lynslager. This document had been created on May 23, 1854 the day before the *Forerunner* left Portsmouth. The document noted that Dr. Brown had been replaced by Heinrich Bleek, and Lieutenant MacLeod by Captain Taylor. However, upon examination, it was found that no provision had been made in the event of Beecroft's death.[99]

Considering the transfer of MacLeod and Brown, and the consul's death, an argument could be made that Baikie was now the senior member of the party. It was apparent that this was what Consul Lynslager believed. Baikie understood the responsibility he was being asked to undertake and believed that the expedition would move in a very different direction under such a junior officer as himself. He argued that he was new to command, unfamiliar with the climate and culture and had never been on the River Niger. Lynslager made it clear that if Baikie rejected the command the voyage would need to be postponed until the following year. However, with the additional information now provided about Heinrich Barth, Lynslager also stated that in his opinion delaying was not an option. No direct order was given. It was clear what Consul Lynslager wanted to have happen, but the final decision would be Baikie's. The brief conversation was over, Baikie accepted command. In his own words, "I resolved to continue the expedition, as I considered that, the preparations being so far advanced, and results of no little importance being expected, it would be wrong not to make the attempt."[100]

ARRIVAL OF THE PLEIAD

They were still sitting in Lynslager's office when one of the servants burst into the room to announce that a ship's flare had been spotted on the horizon. Everyone rushed to the terrace outside the office just as a second rocket was fired. The ship's blue lights could be seen, and an answering rocket was fired from the beach below. One of the seamen at the water's edge quickly hoisted a blue light onto the mast anchored near the shore to guide the ship to the proper anchorage. Through the night glasses Baikie could see that it was a steamer, but it was an hour before it could be verified as the *Pleiad* and another hour before the ship was safely at anchor (Figure 6.2).

Baikie and Lynslager moved to the beach, secured a boat and rowed to the ship. Captain Taylor's only response to the reason for his four-day delay was that there had been a setback on the coast. Baikie was torn between his relief at seeing the ship safely in port, with a full complement of Kru sailors to serve on the voyage, and his concerns about Taylor's aloofness, which seemed to border on hostility. As Baikie and Lynslager were preparing to leave the *Pleiad* they were approached by a tall, distinguished-looking man who introduced himself as Dr. Thomas Hutchinson. The doctor announced that he had been assigned to the expedition as an assistant surgeon and that he was to report directly to Baikie. Hutchinson had traveled with Taylor on the *Pleiad* from Dublin, yet in their previous meeting Taylor had apparently not thought it important to mention this fact to Baikie. Had Beecroft lived, Baikie's job would have been to avoid Taylor as much as possible, draw his charts and maps, and collect his specimens as assigned. But as the new commander of the expedition, Taylor had now become Baikie's responsibility.

Several days later, Baikie stood on the terrace of the governor's mansion and watched the sun slowly emerge from the Bight of Biafra. Soon the terrace and residence were bathed in soft tropical light. Baikie turned and walked back into the consulate. He was worried. Both the Niger and the Benue are greatly affected by rainfall. At high water, the rivers form wide navigable waterways, several miles in width. But in the dry season they shrink considerably, exposing large sandbars which divide the shrunken rivers into several unnavigable channels. The ease of navigation on the Niger would be dictated by the depth of the river, which in turn was controlled by the seasonal rains. The expedition was now weeks behind schedule and falling water levels would continue to diminish the amount of time they would have on the river.[101]

Baikie moved back outside and took a seat along the low wall surrounding the terrace. To the east the calm waters of the Bight of Biafra stretched away from his temporary home on Fernando Po. Somewhere beyond the horizon was the West African coast. Baikie must have imagined a thousand times sailing off toward the Cameroons. When they sighted land, just past the Rio del Rey, they would turn north. They would continue past the mouth of the Calabar River and on to the River Brass. A little farther along the coast, just beyond Palm

Figure 6.2 The Pleiad © *Frank Cass and Taylor and Francis*

Point, the River Nun emerged as one of the hundreds of rivers that emptied into the Bight. This would be the initial goal of the expedition. This was the passage that would lead to the main channel of the River Niger.

Beyond the terrace wall the hill fell away, stretching across the narrow beach to Clarence Cove below. There the *Pleiad* lay at anchor. From this distance it was difficult to determine if there was any activity onboard yet. Baikie stepped to the stairs just beyond the edge of the terrace and began his descent from the top of the cliff to the bay and the *Pleiad* below.

Although the ship had arrived late, it had seemed wonderfully well designed for the challenge. But except for having a good ship, little had gone right for Baikie since the beginning of the voyage. First, there was the replacement of the captain and chief surgeon. This was followed by the death of Beecroft two weeks before Baikie's arrival. Bleek, the linguist, had been added, with no role to play and now his worsening physical condition was leading Baikie to question Bleek's ability to continue the voyage in any capacity. The unexpected arrival of Dr. Hutchinson would help alleviate Baikie's medical responsibilities. However, Baikie was still the physician in charge and now would also be leading the expedition. Due to this series of isolated yet related events Baikie had gone from fourth or fifth in command to heading the expedition. But the appointment of Taylor as captain was proving to be the most problematic development of all as he was proving very difficult to work with.

As Baikie reached the shore, he unhooked the lines on one of the boats used to take crew and passengers to and from the ships at anchor. He launched the boat, pushing out through the waves heading toward the *Pleiad*. As he climbed over the side of the ship, he could hear the sounds of stores being put away and last-minute preparations being completed. But it wasn't the crew of the *Pleiad* that was completing these tasks. The sounds of work were coming from the crew of the *Crane,* another British ship at anchor at Fernando Po. Captain Taylor was still absent and continuing to provide no direction.

CAPTAIN TAYLOR

Although Baikie had been selected to assume Consul Beecroft's duties he had neither responsibility for, nor experience in preparing a ship ahead of a voyage. The responsibility fell to Taylor, whose erratic behavior was putting the launching of the expedition even farther behind schedule. Taylor was proving to be completely inept at command. His first task upon arrival at Fernando Po had been to ready the ship and to secure stores and coal on board. Although he was supposedly an experienced captain who would have assumed this task many times in the past, it seemed that Taylor was entirely unsuited for this work. He was late getting up, later getting started and quit before he could get too tired. Things were moving in an irregular manner, with stores being randomly loaded. On some days the loading was completely neglected, and on other day's goods were loaded with great haste. Things were often put on board so quickly that the ship had to then be unloaded the following day and repacked. There was also unrest among the ship's crew.

The job of the supercargo, who is responsible for overseeing the cargo, is of key importance on a vessel engaged in commerce. His duties include managing all the goods that had been put onboard to be used for trade. He was also responsible for receiving goods collected from trade while in Africa to be carried back to England on the return voyage. Macgregor Laird had financed this voyage with the expectation of turning a profit through his involvement. The *Pleiad's* supercargo, a most critical role, had been selected personally by Laird but now trouble was brewing. Because of Taylor's carelessness in loading and sorting the ship, he and the supercargo were constantly at odds.

Matters finally came to a head and the supercargo refused to follow any further directions from Captain Taylor, left the ship and boarded a return vessel bound for England.[102] This position was critical on the ship and could not go unfilled. Baikie was forced to promote Samuel Crawford from mate to supercargo. Once again, a position was being filled with a last-minute replacement and with someone who had not been trained for the position.

When Captain Taylor was selected to replace Lieutenant MacLeod, Macgregor Laird had given him a copy of the contract with the Admiralty, along with very detailed instructions as to how he was to carry out his responsibilities. The final requirement stated he was to get the *Pleiad* ready to sail and be underway as early as possible, but under no circumstances was he to delay beyond July 1, 1854. Yet it was already July 3 and it appeared that the *Pleiad* was still days away from being ready to sail.

The *Crane* had lain at anchor near the *Pleiad*. Captain Thomas Miller, who was in charge of the *Crane,* saw the lack of organization onboard the *Pleiad* and delayed his return to England by a week to lend assistance.[103] Each day he would send a crew from his ship to assist in the preparations. Without this help, the *Pleiad* might never have been ready to leave

Fernando Po. This day was no exception. Again, it was the *Crane's* crew that was making the *Pleiad* ready.

As Baikie boarded the *Pleiad*, he observed Daniel John May, second master of the *Crane*, supervising the final preparations. When Captain Miller had offered the services of his loading party it was May who had been sent to supervise the work. Baikie had observed that May and Taylor could not have been more different in character. Where Taylor appeared to be a disorganized slow-starter, May was energetic, reliable, and quick to finish tasks. Now, as they prepared for departure, it was May who again had the *Crane's* crew moving swiftly.

May detached himself from from the group of workers and approached Baikie. He was aware that Baikie had now been named to lead the expedition. The island was not large and few secrets could be kept on it. Now, May made Baikie an offer that had the effect of eliminating many of Baikie's concerns over the upcoming voyage. May stated that he had thought long about the voyage of the *Pleiad* that lay ahead, and that if Baikie had no objections, he would like to accompany them as second mate on the journey. Baikie quickly agreed.

As Baikie was preparing to leave the ship, Dr. Bleek emerged from below deck. Baikie steered him to the edge of the wheelhouse and out of the sunlight. Bleek made no protest as Baikie led him to his cabin. In authority for only a short period of time, Baikie was about to make yet another command decision. Dr. Bleek would be sent home on the *Crane* when she left for England.

THE QUININE EXPERIMENT

It was July 7 and the *Pleiad* was in the final stages of preparation for the voyage. Two large iron trade canoes had been positioned, one on each side of the *Pleiad,* and were now securely lashed to the vessel. It was late afternoon, and the crew was assembled on the deck for a final meeting. There were twelve Europeans and fifty-three Africans all waiting for their new commander to address them. As Baikie faced the assembled group he noted that Taylor was missing. The departure could not be delayed, or they would miss the tide. Postponement was not an option, and he began explaining the three-part mission to the crew as outlined in the Admiralty Instructions.

First, they would chart the course of the River Niger to the confluence with the River Benue. They would then sail as far as possible up the Benue, charting that river as well. Second, while moving toward the Benue, they would attempt to find Heinrich Barth and, should he desire, offer him safe transportation back to Fernando Po and on to England. Finally, they would locate sites for trading enterprises along the course of both rivers that would be established by others on subsequent voyages.

However, it was a final paragraph in these instructions that would give Dr. Baikie the opportunity to make this voyage unique above all others. Beecroft's directions on his responsibility toward the health of those traveling upriver had been made very clear:

> Finally, you are strictly enjoined to be careful of the health of the
> party entrusted to your charge, and to afford them the benefit of

> your [Beecroft's] experience as to the best mode of maintaining
> health in the African rivers; and should, unfortunately, fever break
> out and assume a threatening appearance, you are to remember
> that you are not called upon to persevere in the ascent of the river,
> but that your first care is the safety of your people.[104]

Baikie intended to make this Admiralty reminder the fourth primary goal of the voyage and would attempt to safeguard the health of the crew in a way that Beecroft could not have envisioned. Baikie explained to those assembled that during past expeditions the greatest concern had not been the rapids, wild animals or hostile encounters with the local populations. The greatest concern, especially to the Europeans, had been the fever.

MALARIA

There are certain diseases that have been shrouded in mystery over history; their causes attributed to magical or spiritual forces. Malaria has long been a member of this fascinating group of diseases with veiled origins. For many centuries it was believed that malaria was caused by miasma (an ancient Greek word meaning pollution or defilement), a poisonous vapor or mist filled with particles from decomposed matter (miasmata). It would be decades before scientists would determine that it was mosquitos, not particles in the mist, that carried the disease. References to periodic fever from malaria are found throughout history. By the time of Baikie's voyage people had battled against malaria for hundreds, if not thousands of years. Those affected by the disease knew the locations where malaria was most prevalent, but still had no understanding of what caused the fever.[105] It was generally agreed that something present in the swamp mists caused malaria and those who had studied this issue had formulated several theories on both its prevention and its cure.

From the earliest days, the link between swampy ground and contracting malaria was recognized. The term "malaria" originates from medieval Italian (*mala aria*) literally meaning bad air. It was believed that vapors entered through the skin or lungs and enabled the disease to be transmitted to humans. Varro, a Roman gentleman-farmer, is credited with first suspecting that the linkage might be from insects or microscopic animals rather than from the air itself. Just before the birth of Christ he wrote what amounts to the first treatise on country life, *De Re Rustica*, in which he says of malaria "that in marshes there are animals too small to be seen, but which enter the mouth and nose and cause troublesome diseases."[106]

Baikie was aware that the death rate among Europeans traveling into the interior in this part of Africa had often exceeded seventy percent. According to Liebowitz (1999), malaria was responsible for most deaths throughout Africa among the explorers, traders, soldiers, missionaries, and later colonists. He reports on Livingstone's and Kirk's descriptions of malarial attacks in his book, *The Physician and the Slave Trade* thus: "At first there was a feeling of profound lassitude followed by high fever, shaking chills and muscle spasms. Drenching sweats are followed frequently by a subnormal temperature and profound, lasting weakness."[107] On Clapperton's second journey it was recorded that Richard Lander contracted malaria and suffered from such debilitating muscle spasms and such "agonizing pain in his limbs" that he could ride for only a few miles without dismounting from his horse and "rolling on the earth in the hope of relieving the pain."[108]

CINCHONA TO QUININE

At the time of Baikie's voyage a London physician was still proposing that Europeans could protect themselves from the mists if they changed their bathing habits by oiling the skin instead of using soap and water:

> I may state that one of the conclusions at which I arrived as the result of some very extended enquiries into the nature of fever was that what we call malaria makes its noxious impression, not upon the lungs, but upon the general surface of the body. It was this view which led me to entertain the idea of the possibility of defending the skin from the action of malaria by means of some unctuous application, or oil alone. There is a very remarkable circumstance that the most distinctive characteristic in the personal habits of the natives of Africa, as contrasted with those of strangers who visit them, is that the common custom of the one people is to anoint the whole surface of their body freely, while the other, on the contrary, with the aid of soap, are in great pains to remove everything of the kind which even the natural recreation of the skin provides. Here, then, is a broad distinction between the personal habits of the two people—the one anoints, the other washes. They are both equally exposed to the influence of malaria—the one escapes, the other is annihilated. I would only add that I think it important that it should be used in the evening as well as in the morning, because it would appear that the influence of malaria is most powerful between the hours of sunset and sunrise.

Baikie knew that the use of oil on the skin had proven ineffective; however, he agreed that malaria was caused by something in the swamp mists. There were two common practices that had proven marginally effective in preventing and treating fever. The first practice involved keeping the crew members away from the mist. This was the reason that most vessels remained anchored offshore, using local inhabitants to bring the products from the interior of the country to the ships. When Laird had traveled upriver, he had awnings stretched over the deck to protect his men from the ill effects of the dew and provides the following insight: "If certain simple precautions were more rigorously observed, such as never sleeping on shore, or reducing boat work up the creeks to a minimum, the chance of infection would be much diminished."

The second practice involved treating those who had contracted the disease. Most of the Jesuits trained in Rome had seen malaria victims and were familiar with the disease. In Peru, in about 1630, a Jesuit Brother named Agostino Salumbrino wrote that a cure for malaria had finally been found. Brother Salumbrino was an apothecary by training and had observed the effect of quina tea, extracted by the Quechua Indians from tree bark, in treating malaria. Salumbrino described collecting and drying the quina bark. After drying, it was ground into a fine power and mixed with water to form a strong tea that was then consumed. The only modification that the Jesuits made to the traditional cure was to mix the powered bark into wine instead of water.[109]

There is another story of its origin. During the seventeenth century Peru was governed by a viceroy, the Count of Chinchon. In this story his wife, the Countess of Chinchon, contracted malaria and was successfully treated by the Jesuit Brothers. Having discovered the properties of the quina tea, she worked with the Jesuits to identify the bark from which the tea was brewed and carried some of it with her on her return to Europe, presenting this gift of life to the Pope and thus introducing the cure for malaria to the European world.

This story was recounted by Sebastiano Baldo, a Genoese physician. Based entirely on this account, the seventeenth century Swedish botanist Carolus Lannaeus named the plant after the countess (Cinchona). It is a wonderful story and virtually none of it is true. Not a single document has been discovered to support it and the account appears to be a work of pure fiction concocted by Baldo. The official diary of the Count of Chinchon contradicts all such claims. It states that Ana de Osorio, the first Countess of Chinchon, died in Spain three years before Philip IV appointed the Count to be Viceroy of Peru. The second Countess, Francisca Henriquez de Ribera, accompanied her husband to South America. While Dona Francisca continued to enjoy excellent health, the Count had several episodes of fever, none of which were treated by the bark. It is also recorded in the diary that the second Countess never returned to Spain; dying in the port of Cartagena, Columbia during the trip home. Thus, although it was the first legitimate cure for malaria, the arrival of cinchona bark in Europe was entirely unrelated to the Countess.[110] Following the discovery by Brother Salumbrino, using the bark of the cinchona tree for treating fever had quickly been adopted by the crews of slave ships traveling regularly between Europe, Africa, and South America. The inclusion of cinchona bark in the *British Pharmacopoeia* of 1677 put it into the class of respectable medicines. Thereafter, for nearly two hundred years, the bark was employed in great quantities as a cure for malaria with no-one knowing why it was successful.

Royal Naval surgeons had been using quinine for years to treat sailors who had contracted malarial fever. However, this treatment evolved through observation of the slaving crews and was not something that the surgeons had learned at university; and for reasons that are largely unclear, the use of blood-letting and purgatives remained the standard method of treating malaria by the general medical establishment. Thus, a leading physician had written in 1850:

The mortality of Europeans on the coast is reckoned at anything up to 700 per thousand per annum. The poisonous exhalations causing malaria can be found arising in the morning mists from the humid mangrove swamps of the delta, smelling of death and putrefaction. Bloodletting and mercury treatments seem to have the greatest impact on the disease, fifty ounces of blood at the onset of the illness, or fifty grains of calomel a day to purge and dehydrate is the normal practice. [111]

Science finally isolated the quinine in cinchona through the work of two French chemists, Pierre Pelletier and Joseph Caventou, in 1817. These men elected to ignore the story of the Countess and instead chose the original quina tree to name their extract. Thus, cinchona became quinine; but it was still used as a treatment rather than a preventative. It was not until about 1850 that quinine was proposed as a prophylactic in preventing malaria. To do so was suggested in the writings of the naval surgeon Alexander Bryson.

BAIKIE AND BRYSON

Dr. Alexander Bryson (1802–1869) began his career in the Royal Navy as an as assistant surgeon in the West Africa Squadron. In 1847, Bryson presented the Admiralty with the "Report on the Climate and Principal Diseases of the African Station" in which he announced that by using quinine the annual mortality rate had been halved. He recirculated it two years later to a wider audience as "An Account of the Origin, Spread & Decline of the Epidemic Fevers of Sierra Leone."[112] Bryson blamed the medical schools for neglecting this mode of treatment in favor of traditional blood-letting and purgatives. He is also credited with first mentioning the use of quinine as a prophylactic. Somewhat brusquely, he indicates that his fellow naval surgeons knew more about conditions in the tropics than all the academics in London and Edinburgh. He adds:

Cinchona bark and the sulphate of quinine are both extremely useful agents for the prevention of fever; and although their administration is apprehended but indifferently understood, still the numerous instances on record in which they have been successfully employed leave no doubt that their more general use upon the station is most urgently needed. It is firmly believed that although neither bark nor quinine has the power of preventing the germs of fever from lodging in the system, nevertheless they most decidedly have the power of preventing their development in pyrexical action.

Baikie was convinced that Bryson was correct in his assumption that the regular use of a consistent dosage of sulphate of quinine before one became ill could serve as a prophylactic to prevent high fever. When taken with wine this medicine had proven a great success in treating those suffering from fever. Baikie told the crew members that it was his belief that this treatment would also prevent them from contracting malaria and yellow fever if the procedure was started before one became ill. Therefore, starting that day, every person would be given two-thirds of a glass of wine containing five grains of quinine each morning and a second glass before retiring in the evening. With that, the three mates began distributing the glasses to each European member of the party.

Baikie was now prepared to stake his reputation and the lives of those under his command that his fellow Royal Navy surgeon knew more about malaria than his former medical professors at Edinburgh, using what we would today term a clinical trial. If he were correct, the center of Africa could be opened up to outside exploration. If he were wrong, in either his theory or in the distribution of the drug, he could lose his own life and the lives of all those entrusted to him.

As Baikie was preparing to go below, the ship that had been anchored beside them for the past weeks, and whose help had been vital to their preparations, hoisted anchor and began to move from the harbor. Standing at the rail of the *Crane* was a very frail looking Wilhelm Bleek. When Baikie had broken the news to Bleek that he was ordering his return to England there had been disappointment. However, after considering the options and alternatives Bleek had reluctantly agreed. Now, as the ship slowly moved past the *Pleiad* lying at anchor, the doctor gave a slight wave then moved to the hatchway leading to his cabin below deck.[113]

CHAPTER SEVEN

THE KWORA AND TCHADDA

After all the misfortune that had befallen the expedition it appeared things might be turning around. In addition to Daniel May serving in his role as second master, John Dalton was quickly proving himself to be vital member of the crew. With no current responsibilities as zoological assistant he was constantly lending support to other areas of the ship. Baikie had also secured the services of Thomas Richards, a Yoruba man. Richards had served for years as Beecroft's assistant and had accompanied him on many of his earlier explorations. Richards' knowledge of the river would provide the kind of support that had been missing since the Consul's death and could prove crucial in locating Heinrich Barth.

The delay in leaving Fernando Po had also allowed Simon Jonas to catch up with the expedition. Jonas had been with Reverend Crowther on the 1841 voyage but had been away when Crowther had boarded the *Forerunner* at Legos. Both Richards and Jonas would serve as interpreters in addition to their other duties. Baikie had also elevated Dr. Hutchinson from junior surgeon to surgeon in charge of all medical needs. Baikie would maintain the supervision and monitoring of the quinine distribution, but for the remainder of the crew's medical needs Hutchinson would be in charge.

According to Consul Lynslager the time to sail was at high tide and that time was rapidly approaching. Baikie once again checked his watch, as behind him the crew hurried to make a final check of the iron canoes lashed to each side of the *Pleiad*. The canoes, fifty feet in length and eight feet across, were nearly as long as the ship and the sailing master had loaded each of them with coal. Captain Taylor gave the command to cast off lines and the *Pleiad* moved away from her moorings. On the shore Lynslager was backlit by the sun as he alone waved them on their way.

As they passed beyond Cape Bullen, having barely cleared Clarence Bay, the ship was suddenly pitched violently by the fast-running sea. The heavy coal-laden canoes tied to the sides of the *Pleiad* were making it impossible to steer. Taylor ordered that they be cut loose and towed, one on each quarter. May advised against this, saying that if they started toward the ship there would be nothing that could be done to keep them from crashing into the *Pleiad* and punching a hole in the hull.

Taylor looked at May as if to speak but then turned back to the crew and repeated his order to secure tow lines on each quarter. He also ordered eight of the Kru oarsmen into each of the canoes to help steady them with their paddles. This done, the *Pleiad* had preceded less than a mile when the canoe on the starboard side was caught be an unusually large wave. The metal ram, filled with over a ton of coal, hurtled toward the ship. Taylor had just an instant to increase the throttle. This action, plus some furious work by the Kru with their paddles, changed the direction of the projectile just enough for it to pass safely astern. As the canoe went by, traveling at tremendous speed, Baikie could see the fear on the faces of the oarsmen.

Luck and the skill of the Kru had avoided the collision; however, both canoes were now on the same side of the ship creating a momentum that was pulling the *Pleiad* sideways through the water. Crew members frantically pulled at lines and the sailors sitting on top bent into their paddles until the errant vessel was finally back on the starboard quarter. May ordered that both canoes be tied astern and this time Taylor did nothing to interfere.

As the morning was producing a very heavy sea, Mr. May suggested using the sails to steady the ship. At first Taylor seemed not to have heard and May started to repeat the suggestion. Taylor responded first by insisting that the sails were not needed and ultimately admitted that he had decided to leave the sails in the warehouse on Fernando Po.

Baikie's excitement at the launching of this long-awaited adventure was soon replaced by anger and mistrust. The crossing from Fernando Po that should have been accomplished in two days had taken four. The *Pleiad* had almost been sunk by the coal bearing canoe. The lack of sails had continued to cause the ship to drift off course. A steam valve had then locked requiring the engine to be shut down for an additional four hours. This caused the *Pleiad* to drift even farther away from the point where they had hoped to enter the delta that formed at the terminus of the Niger.

THE DELTA

Once they reached the coast, it seemed as if Taylor was unsure of their location. The ship steamed close to shore where they had been forced to battle against a strong current. Upon reaching the River Brass, Taylor had ordered the ship to anchor for the night. Several hours of daylight remained and the charts had shown the Brass and River Nun were less than an hour's sail apart. But Taylor had maintained he was afraid of missing the river with the loss of light.

Finally, on July 12, the *Pleiad* lay at anchor just outside the River Nun and once again they were waiting. All hands were on deck looking toward Palm Point, the landmark indicating the mouth of the tributary that would lead them eventually to the main branch of the Niger River. This great waterway was what the locals called the Kwora. About three miles from the mouth of the River Nun the water suddenly developed a brownish hue. The change in color and the choppy look of the waves could only mean a large sandbar blocked the entrance. The shallowness of the water caused an uneven pattern of the waves that could easily swamp a ship much larger than the *Pleiad*. The shifting sand also made locating the best spot for crossing a skill few sailors possessed. Richards was selected to guide the boat in, being the person most familiar with the approach to the river.

With Richards positioned in the bow to serve as pilot the ship moved forward. The crossing was going smoothly and Baikie must have been thinking that their luck, or Taylor's skill, might have improved. But at that point, the line attached to one of the canoes split and it was cast adrift. To turn and try to retrieve it at this point in the crossing was impossible and all eyes turned to look at the Kru sailors waving franticly. It was soon apparent to them that they were on their own and were once again forced to paddle for their lives. The momentum that had been established and their highly honed skills, employed for the second time in this short voyage, soon carried them past the bar and into calmer waters. The *Pleiad* too made it safely and was soon moored near the shore.

After staying at anchor for several hours and making last minute repairs on the engines they entered the Nun and began searching for the main channel. This proved difficult. It was impossible to find a point of reference along the edge of the river. The darkness created by the wall of trees that grew to the water's edge meant there was no visible shoreline. Deep water would suddenly be replaced by a sand bar only a few inches below the surface and in the darkness the ship was constantly cast aground. Numerous streams, some wider than the main channel, lured the *Pleiad* up one blind alley after another. Hours were wasted traveling up a wide avenue that suddenly tapered into nothing. The return trip to the original channel proved equally problematic as a series of seemingly identical waterways crisscrossed dense lines of mangrove trees that extended to sixty feet in the air, leaving only a thin slit of sunlight overhead.

It was decided that Baikie and Richards would leave each morning in one of the small boats and would spend eight to nine hours searching the next section of the river. After an exhausting time, they would return to the ship with difficulty, retracing their route though the repetitive landscape of the delta. The following morning, they would move the *Pleiad* forward with the first light. When the ship reached the point that Baikie and Richards had explored and marked the day before they would drop anchor until the next stretch of the river could be searched using the small boat. Day after day this exercise was repeated. However, even with the course clearly marked, Taylor seemed unable to keep the ship from being repeatedly grounded each day.

It was several days before the *Pleiad* could clear the delta. But eventually the swamp was replaced by the powerful Niger River that at times reached 200 yards in width, its current swollen by the previous six weeks of heavy rains. The delta had been bleak and uninhabited. The mangroves, packed close together and continually under water, formed a natural barrier that prevented any kind of dwelling. This natural barrier had prevented any kind of human habitation. Upon leaving the delta and reaching the main channel of the river, the dense mangroves quickly gave place to towering palms sitting atop a vibrant green riverbank. Now individual houses and even small villages began to appear as the *Pleiad* continued to push upriver against the strong current. Although the inhabitants could often be seen along the banks and occasionally moving toward the *Pleiad* in their canoes, none could be enticed to come aboard the ship.

SETTING THE STAGE FOR GOD AND TRADE

Few Europeans had ever ventured even this far on the river. Landmarks needed to be identified and charted to create useable maps for those who would follow. Baikie assumed overall responsibility for the charts, with May drawing the actual maps. The task of creating these charts and maps appealed to Baikie and he took the opportunity to name virtually every landmark he encountered. The names of members of the British royal family, prominent politicians and sponsors of the voyage were given to islands, mountains, and streams. Members of his family, friends, and faculty members from his university days were also recognized, honored, and guaranteed immortality along the course of the *Pleiad's* travels. Baikie was also generous in recognizing members of the expedition. As a result, Richard's Creek, May's Point and Taylor's Creek all soon found their way onto the official charts of the voyage. Even the *Pleiad* had an island named in its honor. Baikie bestowed these honors on everyone but himself: not one landmark along the course of the voyage bears the name Baikie.

Baikie and Richards, continuing their advanced exploration in the small boats, were now joined by May. Often, they would land near a village where they would trade goods for food and supplies or to purchase wood to fire the boilers. On these occasions, Baikie would take the opportunity to collect specimens of plant and animal life, which he would then preserve, meticulously cataloging and storing each one carefully in the large wooden chests he had brought along for this purpose. Baikie also recorded sightings of animals and plants too large to transport home. The river itself provided sightings of hippos and crocodiles. Along the banks leopards, hyenas, and a wide variety of hoofed stock could be seen moving along the river to hunt or drink. Higher up on the bank herds of elephants, sometimes numbering in the hundreds, were often in view. He had soon filled half his journals with descriptions of rare plants and animals that he had either collected or sighted. Every day brought a new discovery and not all involved the flora and fauna.

As the ship moved through the various sections of the river, the explorers soon became aware that they were traveling through a series of individual territories, each under the authority of a local ruler. Early in their voyage the resident populations had fled from the party from the *Pleiad* as they attempted to make contact. As the ship proceeded further upriver, however, rather than fleeing at their approach, canoes began drawing near the ship with an invitation for the white men to pay their respects to the local leader. Even if they were pressed for time the offer was never refused for three very specific reasons: First, the primary aspect of the mission was to determine prospective locations for future trading sites. This required an assessment of the local ruler's power, influence, and receptivity. It also involved a careful inspection of the village that could not be adequately accomplished from the water. Second, it was hoped that the people they encountered would have information on Barth's location, as his travel toward the coast would have undoubtedly followed the river. Finally, Richards had told the party that those in the past who had traveled through a territory without acknowledging their host had sometimes paid with their lives. So, perhaps the primary motive in accepting a local leader's hospitality was to assure the group's survival.

At each meeting Baikie, Crowther, Richards, and May would enter the village and would be directed to the dwelling of the head man. On this section of the river there was no need for a translator, as both Crowther and Richards spoke the language. After paying their respects, gifts would be exchanged. The gifts presented to each host varied from bolts of cloth and mirrors to sabers and muskets. The value of the gifts was based on Baikie's assessment of the degree of power held by the ruler. In return, the visitors would be given goats, cattle, chickens, eggs, fruit, and vegetables. Baikie would ask about Heinrich Barth and any information that might help them pinpoint his current location. Next, Baikie would explain the purpose of his mission and ask for permission to trade with the villagers. There was an assurance that those sites showing the most promise would later be selected for a permanent market center. Finally, Reverend Crowther would be introduced and would ask the host if he would be agreeable to building a school and a church in the village.

The importance of "the missionary impact" on these local rulers was key to the success of the commercial interests that would follow. However, the rulers were faced with a difficult choice. If they admitted Christian missionaries and encouraged missionary activities, they could gain the advantage of schools, technical assistance, military aid, and the development of communication with the centers of Western power. But in return, they had to pay the price of the undermining of the system of beliefs on which their authority was based, and the emergence of a community of Christian converts whose political loyalty would be doubtful.[114] Despite this risk, and without exception, the rulers seemed enthusiastic in their willingness to accept Reverend Crowther's proposal. However, Baikie believed their willingness to embrace the risks of a relationship with the Christian God was directly related to the rewards of a permanent trading post that came with the church.

On July 26 the expedition arrived at the Igbo village of Onitsha and was warmly received by Obi Akazua and his court. The conversation followed the established pattern that the landing party had devised with Crowther ending the meeting by discussing his plans to open a mission and school. As it happened, Onitsha was at that time at war with a neighboring village. Akazua thought a permanent visible sign of his connection to the powerful English might aide him in these efforts; so he readily agreed to supporting the mission and donated a tract of land before the meeting ended. The following week, on August 2, the first church started to be built. The new school also opened, accepting its first twelve students. The word of the mission and school quickly spread. The Igbo spirit of rivalry, competition, and emulation was set in motion and soon requests for schools and missions were coming into Onitsha from villages that the *Pleiad* explorers had not even visited. This would have a far-reaching impact on Baikie's next voyage, which was the founding of Lokoja and the British trading efforts that were to follow well into the next century.[115]

TAYLOR IS REPLACED

The morning of August 5 found the *Pleiad* anchored below the base of Mount Patti at the convergence of two of the largest rivers in West Africa. Here the Benue River, which was known to the locals as the Tchadda, separated from the Niger River, called the Kwora. While the Niger flowed to this point from a northwesterly direction, the Benue approached the confluence from the northeast.

Baikie, Crowther, and May left the ship and climbed up a narrow winding path toward a small plateau. Beneath them was the green-topped Mount Stirling and the heavily wooded tract which had been designated in 1841 for the Model Farm Project. The two rivers lay beneath them. On the left, the narrow channel of the Niger flowed along its meandering route following the base of the western highlands until reaching the confluence. To the right, flowing directly toward them, the broad Benue stretched straight as an arrow for as far as the eye could see. It was in this direction that the next phase of the *Pleiad's* odyssey would occur.

Coming down from the mountain the explorers encountered a large canoe anchored near the shore, its African traders exchanging all kinds of merchandise: ivory, country cloth, tobes, mats, shea butter, palm oil, yams, sheep, goats, and fowl. There was a considerable crowd and everyone was clamoring to buy their goods. The local people bought the goods using cowries, or in exchange for the English goods provided by the crew of the *Pleiad*.[116] Baikie left the crowded scene and crossed the river to walk over a portion of what had been the Model Farm. The shore was lined by a thick belt of trees and beyond this forest grew grass so tall that walking was difficult. Much of the land was low and swampy. However, there were also elevated spots of projected rock and large sections of choice land upon which crops could be easily grown. The entire area was abandoned, with the only signs of life being the large insect nests and tracks made by antelope and elephant. But in fewer than ten years this would become a thriving trade center called Lokoja and Baikie would be known as its king.

Leaving the confluence brought further delays. Although wood was abundantly available near the Model Farm site only a little had been taken on board. Taylor had decided that wood could be obtained along the course of the Benue. Yet what little they could find was of poor quality and did not burn well. This meant stopping five or six times a day to gather fuel. When the ship was underway the wood generated so little heat that they would often feel they were standing still against the sharp current. The result was that the *Pleiad* was unable to travel more than six or seven miles in a day. The delay in starting from Fernando Po, the extended period taken in reaching the confluence and now the frequent stops had put the expedition weeks behind schedule.

The following day was no better, and Taylor seemed to be have no solution. Baikie ordered a party to travel farther inland in search of wood and a village was located within a mile of the river. At the expense of a few hundred needles and some little zinc-cased looking glasses they were able to fill one of the iron canoes with wood of excellent quality already cut to length. A steady wind blew upriver and Baikie took the opportunity to create masts from the wooden poles collected at the village and sails from the pieces of canvas that had covered the trade goods. He then attached these onto the two canoes. With the canoes tied to each side of the *Pleiad,* the steady wind, and the better burning wood, they took the ship forward at excellent speed. They covered fifteen miles, almost three times the distance they had previously covered in a single day.

On the afternoon of August 16, the ship once more ran aground. Since leaving Fernando Po Taylor had not once left the ship. He had shown no interest in the trading ventures, meeting the local rulers or in planning and evaluating the course that the ship would take. However, on this day he announced his intention to take one of the small boats and to determine the river depth himself. He was gone less than two hours and upon his return continued shaking his head gloomily. He indicated that the ship was "ominously placed," and that they would probably be on this sand bar for several days. Taylor also stated that the immediate route ahead looked no better. He announced he would travel farther the next day.

The following morning Taylor left early and did not return until sundown. Upon his return he quickly grabbed something to eat and drink and then asked to speak to Baikie. He said that he considered it impossible for the ship to advance further. There was no channel, while there were large sandbars and little opportunities for trade. In fact, Taylor said, he thought the river was at an end and it was about to empty into a large lake. Baikie declared that he disagreed on all aspects of Taylor's assessment. In his surveys he had consistently found a channel of at least four fathoms (24 feet), the bars could be avoided, and the villages seemed to be growing more numerous and larger. Finally, Baikie reminded Taylor of the party's climb to the elevation above the confluence. They had seen for miles across the landscape from that height and could see nothing but river as far as the horizon. It was not plausible that they were now entering a lake that formed the origin of the Benue.

The exchange became louder but came to complete silence when Baikie announced that he was relieving Taylor of command and would take the vessel upriver himself. Taylor, taken by surprise, attempted to modify his position. However, Baikie would not back down. Mr. May and the other officers were asked to join the discussion. Except for support from one of the junior officers, none of them backed Taylor. Taylor then began to talk of "mutiny and piracy." But, seeing he had no support for that stance either, he retired to his cabin with its couch and cigars. He was seldom seen again for the remainder of the voyage.[117]

Baikie's first order was to promote the chief mate, John Harcus, to sailing master. Harcus would be responsible for navigation and looking after the general wellbeing of the ship. With the issue of Taylor settled, the voyage seemed to slip into a comfortable routine. Meat, fruit, and vegetables were readily available, and the local people seemed eager to trade. The ship's progress was steady under the new sailing master's guidance and the *Pleiad* was seldom aground for the remainder of the voyage.

THE SEARCH FOR BARTH

While dealing with his charts and maps and concentrating on trade, Baikie was also always focused on gathering information on Barth. Eduard Vogel, like Barth, was a German explorer working with the Geographical Society. He had been selected by the British government to join the Richardson, Overweg, and Barth expedition and to provide them with supplies.

Vogel traveled by ship to Tripoli where he attempted to catch up with Barth. After missing each other several times, the two finally did make contact in Bornu in December 1854, where Vogel discovered that Richardson and Overweg had died weeks earlier. There was almost constant friction between Vogel and Barth, and they had quickly parted ways. There was no way Baikie could have known of this separation and he assumed that Vogel and Barth were still traveling together. Thus, it was no surprise that while he was visiting the village of Ojogo, Baikie was told by a villager that two white travelers had visited only six days previously. They were now believed to be nearby in a village called Keana. The only concern was the time that the *Pleiad* had now spent on the Benue and the report received while on Fernando Po had said Barth had been sighted on the upper Niger River. This would mean that Barth was traveling away from his suspected route, or that one of the two sightings had to be in error. Baikie needed to be sure.

Baikie showed the villager engravings of Barth and Vogel. The man immediately identified Barth, who he said now sported a heavy beard. He was less sure of the second traveler, although he said he looked like Vogel. This was enough for Baikie. He immediately asked a guide to take him and a small group to the neighboring village. Baikie's plans were dashed when he was told that because of fighting in the area it would be too dangerous for his party to travel. However, the townspeople would send a messenger, bearing a letter from Baikie, to notify Barth of his pending rescue.

Baikie waited for eight days, receiving no word from the messengers. In desperation, he again carefully questioned the villager who had delivered the initial information. It was discovered that a major error in translation had occurred. The white travelers had been sighted, not six days, but six weeks earlier. Baikie now understood. Barth had been in Ojogo and Keana before being sighted on the upper Niger. He was traveling toward the coast as suspected; however, this meant that he had already passed through this section of the Benue well ahead of the *Pleiad's* arrival.

Without the expedition's late start and constant delays, they might have been positioned to make contact with Barth. Now it was clear that the expedition would not be able to fulfill this part of their mission. Baikie was distressed at missing the opportunity to carry out a primary part of his work. But, still more upset at having wasted over a week that could have been spent in trade and exploration, Baikie gave the order to resume the voyage.[118]

Although the Model Farm Project had been abandoned, the site where the Niger and Benue rivers merge could still be a prime location for commerce. The location allowed those coming from the north to trade at the confluence rather than having to travel all the way to the delta (Figure 7.1, *see colour section*). To Baikie this seemed like the logical place for a temporary trading site to be established. He decided to send a party back to the confluence to trade until the *Pleiad's* return. In preparation he had one of the canoes cleaned out and stocked with trade goods. The next morning Baikie selected eight Kru sailors and three of the translators for the mission. The supercargo, Samuel Crawford, was selected to head the trading party and he would be assisted by the engineer, Richard Gower. The canoe was soon dispatched for the manageable trip back down river and the *Pleiad,* with its reduced numbers, continued its journey up the Benue.

SUCCESS AT TRADE

Trading continued for those remaining on the *Pleiad* as well. The farther they traveled along the river, the larger was the supply of elephant tusks and hippo teeth that were available for trade. At the end of a single day the group determined they had purchased 620 pounds of ivory. The trade for this valuable commodity had been made exclusively for pieces of white calico and linen handkerchiefs. Palm oil and tobacco and were also in abundant supply from the local people.

As the ship proceeded upriver, they were also able to collect large quantities of shea butter. This white fat, which is extracted from the nut of the shea tree, was widely used in soaps and cosmetics. When it was transported to England its value was greater than that of the palm oil and it was easily collected in trade for a minimum amount of manufactured goods. Baikie also bought many ornaments as representative cultural objects, including bracelets and rings made of copper and brass. These he secured in his storage chests along with the botanical and zoological specimens.

As the ship entered the upper Benue the stands of trees were surrounded by vast expanses of grassy savannah and here they entered the land of the Hausa. Visits to local rulers continued, and on one trip they were promised three horses to assist the men in their return to the ship, although when it was time to leave, only one horse was delivered. William Guthrie, as the oldest of the three travelers, was given the horse. Hutchinson and Baikie began the long journey back to the ship on foot. After about a mile Dr. Hutchinson decided to return to the village to secure the other two horses. Baikie indicated he would walk ahead until Hutchinson could catch up.

Baikie walked alone for about seven miles when he lost all trace of the path. Using his compass, he attempted to work toward the river but became even more entangled in the thick brush. Periodically he would climb a tree to try and get his bearings, but he could not find a single guiding landmark. It was now past sunset, and the darkness was quickly falling around him. Baikie checked his watch and saw it was after seven pm. After attempting without success to continue his search, he found a large baobab tree with a double trunk. With his back against one trunk and his feet against the other he climbed up to a hefty branch about fifteen feet off the ground. Using a piece of cord, he tied his arm around the branch to steady himself. It now dark and he quickly fell asleep.

After about four hours Baikie awoke to the buzz of mosquitoes assaulting him on all sides. Silhouetted against the star-filled sky were several large birds roosting in the tree above his head. His back was stiff from keeping in the same position for so long and he began to climb down. He had moved only slightly when a high-pitched howl erupted from the base of the tree. These were the unmistakable high notes of a hyena. Although he could not see into the darkness, Baikie could hear the hyenas ran off into the brush and a few minutes later he again began his decent. His relief was short lived, as the yap of the hyena had been replaced by the deep bass growl of a leopard. Baikie climbed back up to his perch and wondered if he would be able to see the leopard as he climbed upward from the deep grass below. Some time later he fell asleep again. When he awoke the first light of morning was spreading through the tops of the tree. Baikie's night visitors were nowhere to be seen and he gingerly climbed down from his perch.

Torn between walking back toward the village and walking forward to where he believed he would find the river, he elected to move forward. Baikie had traveled about three miles along a gradually widening path when he came to a small village. Although they were initially surprised and frightened, the villagers soon greeted him warmly and offered him food and drink. At around midmorning two Kru sailors from the *Pleiad* slowly entered the village. Their shock at seeing Baikie could not be hidden. They refused to answer any questions until they were on the walk back to the ship. After traveling several miles toward the river, they finally admitted to having been sent to locate and recover Baikie's body. Everyone had assumed that he could not possibly have survived the night.[119]

CHAPTER EIGHT

THE HERO COMES HOME

T he *Pleiad* had been on the river for over a hundred days. Although the daily progress varied greatly, they had been able to average a good speed. The charts, notes, documents, and maps created by Baikie indicate that they had penetrated over 600 miles into the center of the continent. That meant they had also explored and charted over 250 miles of the Benue River beyond the farthest point reached by any previous European explorer.

THE DULTI

The ship was now entering the area of the river controlled by a group called the Dulti. For days villagers in the area had warned them to turn back before entering Dulti lands. These people were described as primitive, savage, and hostile. Based on these reports, the decision was made to leave the *Pleiad* anchored in the middle of the channel while Baikie and May selected ten of the Kru crew to row, as they scouted ahead using the remaining cargo canoe. At mid-morning, while attempting to locate wood to burn in the *Pleiad's* boilers, a village was sighted. As it was now nearing the end of the rainy season, the huts were surrounded by water from the swollen river, allowing the canoe to move silently through the village. There was no sign of any inhabitant and the village seemed to be deserted. Standing on a slightly raised patch of dry land near the center of the settlement was a large tree. Baikie directed the rowers toward the site and the bow of the canoe soon was resting against the massive trunk.

Baikie and May immediately set up the tripod and began surveying to pinpoint and chart their location. The Kru sailors started the task of breaking off the lower branches of the tree and cutting them into useable sections of firewood. Baikie and May had nearly completed their work when it became apparent that the village was far from abandoned. From all sides waves of warriors had seemingly emerged from the water and were silently moving forward. Half-wading and half-swimming, the intimidating hoard pressed ever forward toward the canoe.

As the Dulti began to draw closer it was clear that they were well armed. It was also obvious that the women and children were being sent away. There were now nearly four hundred armed fighters within fifty feet of the grounded canoe. Some were close enough to grab hold of the vessel and they began probing the bags of trade goods for things that could be easily taken away. Baikie seized a few trinkets and, after handing them to those nearest the canoe, quickly shoved off while the warriors were still examining their treasures.

The cargo canoe was soon flying across the flooded plain with a train of Dulti canoes in close pursuit, while three other canoes came from the side to cut them off. The Kru oarsmen worked madly against the current. The pursuers, paddling at full speed just to keep up, were unable to use their weapons. The cargo canoe was two hundred yards from the river and was making for open water via the shortest route possible. Anticipating the worse, Baikie took out his revolver but found that the ramrod was stiff and immoveable. May got a little pocket pistol ready and Baikie, dropping his pistol into the bottom of the canoe, drew his cutlass. An opening in the brush suddenly appeared before them and with a final stroke they shot into the open river. There their larger canoe had the advantage and the Dulti quickly gave up the chase, returning to their homes in the flooded plain.[120]

The swiftly moving current soon carried the cargo canoe back to where the *Pleiad* lay at anchor. Baikie called a meeting of the ship's officers to discuss their options. It was October 5 and the rainy season was clearly coming to an end. Looking toward the banks of the river it was evident that the height of the Benue was dropping rapidly, at a rate of between one to three feet per day. Laird's directions to Captain Taylor had instructed him to "proceed up the Tchadda without delay, making every effort to reach Yola (near the present border with Cameroon), the capital of Adamowa." They had not yet reached this destination and it was unclear how far beyond their present position they would need to travel. Taylor's instructions had also stated that this goal "ought to be done before 1 September." The delays had put them over a month beyond that date and it was decided that the *Pleiad* would begin its return to Fernando Po immediately.

WILLIAM CARLIN

On the way upriver, Baikie had given names to the prominent geographical features that were now committed to his maps and charts. As they moved easily along the Benue, Baikie could recognize Mt. Laird, Crowther Island and Lynslager Point. Mt. Traill, which he had named out of respect for his favorite Edinburgh University professor, slid by on the starboard side. Mt. Katherine, Mt. Eleanor, and Mt. Isabella; named for his sisters and mother, appeared and disappeared from view. The appearance of Mt. Beecroft, a tribute to the man whose death had propelled Baikie to leadership, marked the *Pleiad's* rapid approach to the Niger River.

With just two days sailing to complete before reaching the confluence, the *Pleiad* stopped to take on provisions. The local leader invited Baikie and Crowther to his home while the others loaded the provisions. Once inside, the ruler offered to sell them a large quantity of ivory that he had carefully stacked in the corner of the room. Throughout the negotiations a boy of about ten years of age sat on a small stool near the ivory. At Baikie's direction, Crowther asked if it was his intention to also sell the boy. "Of course not!" was the reply. Baikie continued to press the point. He believed that the original response had been a political answer rather than

Figure 1.2 Tankerness House © Orkney Library and Archive

Figure 4.1 The Albert, Wilberforce, *and* Soudan © *National Maritime Museum*

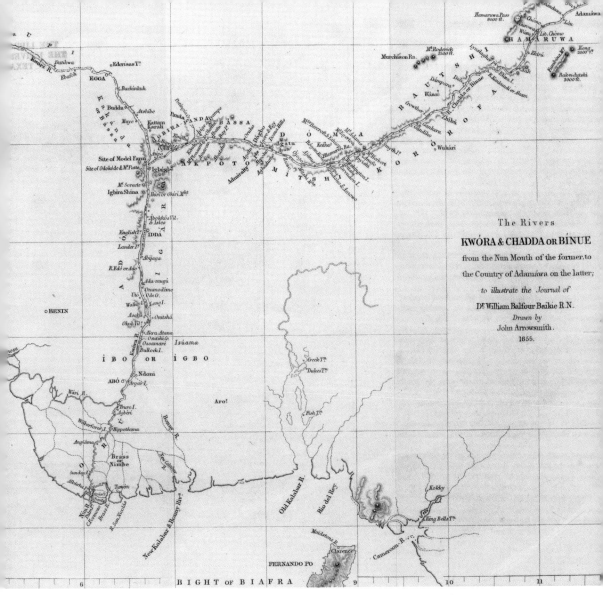

Figure 7.1 Map of Niger and Benue © Frank Cass and Taylor and Francis

Figure 8.1 Front façade of the Royal Hospital Haslar © Haslar Heritage Group

Figure 8.2 (opposite) Baikiaea pluriguga (African teak)
© The Board of Trustees, Royal Botanic Gardens, Kew

57:347

BAIKIAEA PLURIJUGA

1. Young stem showing leaf arrangement, with young leaves.   2. Mature leaves.
3. Flowers and mature fruits.   4. Split mature fruit.

Baikiaea insignis, Benth.

Figure 13.2 (opposite) Baikiaea isignis © *The Board of Trustees, Royal Botanic Gardens, Kew*

Figure 14.1 St. Magnus Cathedral © Orkney Library and Archive

Figure 14.2 Baikie memorial © Orkney Library and Archive

the truth as the ruler had been told when they had visited him on the way up the river that the British were opposed to slavery. Finally, it was admitted that the boy was a slave. Baikie admonished him for lying, but on impulse told the ruler that he wanted to buy the boy's freedom. It was determined that the price for this boy would be 50,000 cowries. It was further determined that this amount equaled £3. Baikie paid in sterling and took charge of the boy. He was not sure why he had done this. He was not sure what he would do next. Perhaps it had been his experience while dining with his cousin at Cape Coast. That dinner now seemed so far in the past. Whatever the reason, Baikie could not simply walk away. The boy looked up at Baikie without expression. He did not understand that he was now free. He only knew that he now belonged to this white man.[121]

Having taken on provisions, the *Pleiad* began its final descent toward the confluence. The realization that they were returning home appeared to buoy the spirits of officers and Kru alike. Even the weather seemed to be cooperating as the daily storms were replaced by calm winds and clear skies. Only the young boy seemed in low spirits. He spent his days isolating himself from others onboard as much as possible. He took his meals alone and slept on deck, seemingly afraid to venture into the confined spaces below. Baikie made sure he was well fed and gave him blankets for the cool nights. Beyond that he left him to adjust at his own pace, content to remain in the open air. Taylor continued his self-imposed exile and never left his cabin. Finally, the forward watch shouted that the Niger had been sighted.

TRADE AND SICKNESS AT THE CONFLUENCE

It had been six weeks since the cargo canoe had been filled with trade items and dispatched to the confluence of the two great rivers. The ivory, tobacco, shea butter, and palm oil gathered by those on board the *Pleiad* now entirely filled the cargo section amidships and most of the storage holds in the bow. Baikie calculated that the products for which they had traded would more than cover Macgregor Laird's costs for supplying the expedition. If the men stationed at the Niger had been equally successful, the voyage would turn a handsome profit as well.

However, when Crawford and Gower were brought on board Baikie was more concerned about their health than the trade goods they had been able to amass. Both men were very pale and quite thin. Their boat had been anchored very close to shore to expedite the trading, but this site had also proven to be quite prone to contracting fever. In addition, although Baikie had sent wine and quinine with them, without his constant intervention the two men had failed to continue the regimen that had been instituted at the beginning of the voyage. Baikie's first action after meeting with his colleagues was to have a meal made of fresh meat and fruit. He then started the treatment of quinine and wine and confined them both to bed rest.

Baikie's next task was to check the results of Crawford's and Gower's six weeks of trading. Because of their illness they had been unable to work for days at a time. As a result, they had not succeeded nearly as well as those on the *Pleiad* had done when traveling upriver. Two hundred seventy-eight pounds of ivory, one hundred and ninety-two pounds of shea butter, an equal amount of tobacco and some red pepper constituted the entire haul. Baikie was disappointed with the lack of trading at the confluence. He maintained his belief that the point where these two great rivers combined was an ideal location for trade and he decided to devote two additional days to trade at their present site.[122]

At exactly two o'clock on the second day at the confluence the whistle sounded sharply to warn the local villagers to leave the ship. In ten minutes' time the *Pleiad* was under full steam and, with a ceremonial firing of the canon, headed down the Niger for the return to Fernando Po. On the deck, Baikie was joined by Crawford and Gower. He was pleased that after two days of rest and treatment they seemed almost back to full health. Baikie was also joined by the former slave boy. Baikie asked Reverend Crowther to explain to the boy that his new Christian name would be William Carlin and that he would travel with Baikie to Sierra Leone, to be educated and cared for. Baikie saw these decisions as a life-changing moment in the boy's life, but the announcement was met only with bewildered silence.

The voyage back down river and the crossing of the bar were uneventful and on November 7 the *Pleiad* entered Clarence Cove on Fernando Po. They had been on the river for 118 days, conducted a successful trading mission covering 600 miles, and explored and charted the Benue River for 250 more miles than any previous expedition. Most importantly, for the first time in the history of African exploration, they had completed the mission without the loss of a single life.

The plan was to leave the trade goods and equipment on the *Pleiad* for the return voyage to England. Baikie, along with his crates of specimens would catch the first available packet steamer bound for Plymouth. The problem with the plan was that next packet vessel would not arrive at Fernando Po for at least two weeks. There was nothing to do but wait. Days were spent completing the journals of the voyage, packing the instruments and collections for shipment, and making clean copies of the charts made from the sightings taken along the Niger and Benue rivers.

THE BACCHANTE

It was not until November 26 that the *Bacchante*, under the command of Captain Dring, arrived from England to take Baikie and the others home. Baikie had anticipated returning to England on the *Forerunner*, the ship that had brought him to Fernando Po in May. However, he discovered from the captain of the *Bacchante* that the *Forerunner* had struck a rock and sunk off the coast of Madeira during the return leg of Baikie's initial voyage. Not only had the ship been lost, but the *Forerunner* had also been carrying all Dr. David Livingstone's journals. The entire summary of his initial expedition on the Zambezi River had gone down with the ship.[123]

After discharging its cargo and taking on fresh supplies, the *Bacchante* was scheduled to leave on November 28. The *Pleiad* would follow the next day under the charge of the *Bacchante's* chief mate. From the time they had landed at Fernando Po the crew of the *Pleiad* had begun to scatter. Some had taken jobs as crew members and officers sailing out to other ports and had already departed on their new ships. Others had agreed to travel with the *Pleiad* on the return trip to England. Some would travel with Baikie on the *Bacchante*. However, few were planning to go as far as Plymouth. As the ship moved across the Bight of Benin and up the coast of West Africa, the number of former shipmates continued to dwindle. The Reverend Crowther and Simon Jonas left at Lagos. The two would travel overland, returning to their duties in Abeokuta. The Kru sailors, so vital to the success of the expedition, returned to their homes at various ports between Cape Coast and Cape Palmas. Soon, the only travelers from the exploring voyage who remained onboard the *Bacchante* were Baikie, May, and the newly named William Carlin.

A transformation had occurred in young William since leaving Fernando Po. He was no longer timid and he no longer avoided contact with the others. In fact, from the bow to the stern, he seemed to be getting into everything. He also had clearly formed a quite close bond with Baikie. Whenever William saw Baikie preparing to go ashore he cleaned himself up and stood by the gangplank, waiting to be invited along. Baikie always allowed him to go. While the ship was underway, he mimicked those who he saw working at their various duties. At times he pretended to be a steward and helped wait on tables. Often, he played at being a sailor. Everyone seemed to enjoy his company and at times the helmsman even allowed him to steer the ship.

By the time the ship reached Sierra Leone the trip was significantly behind schedule and the journey was about to be delayed further. Throughout the voyage from Fernando Po the *Bacchante's* engines had proven highly defective and seemed sadly in need of repair. Top speed could never be attained and travel was painfully slow. As a result, it was already Christmas Day when they reached the port of Freetown. An inspection of the engines resulted in them being declared unsafe and their repair would require an additional ten days.

For Baikie, the ten days were busy and well spent. He made a return visit to Oldfield's warehouse to report on the mission's success and to return the portion of money he had not used for the voyage. He also took the opportunity to make the promised visit to Charles Heddle and his family. Over dinner at Heddle's home, Baikie turned to Charles to secure his support in the supervision of the recently acquired slave boy. Heddle and his family readily agreed to help, and William was put permanently into their care.

With William's future settled, Baikie also took some time in Sierra Leone to relax and to continue working on his notes from the voyage. This much-needed rest and the enjoyable company were obviously valued by Baikie. He wrote to his father, "To Mr. Huddle, I stand especially indebted. His house, during the whole of my time, was my home. A large and airy apartment was set aside for writing in and receiving deputations from the colored population, and all my enquires were most kindly furthered."[124]

Despite his busy schedule and the enjoyable company, Baikie was anxious to return to England. From every point of view the expedition had been a great success. The entire route had been thoroughly mapped and charted. Those people he had encountered were friendly and eager to establish trading relations. The CMS would also be pleased with the progress made. Crowther had already forwarded a letter to the CMS urging the immediate undertaking of a river mission. Baikie was eager to report his scientific findings to the Geographical Society of London and his commercial successes to Macgregor Laird. Yet, it was as if everyone and everything were conspiring to keep him away from England.

One of the responsibilities he completed while waiting at Sierra Leone was his official report to the Secretary of the Admiralty. In this document, Baikie records the personnel he employed upon assuming command after the death of Beecroft. The list included May, Richards, Scott, and Jonas; and he also includes the reasons why each had been employed. With only a passing mention of the problem that leaving the sails in Fernando Po had created for the mission, he omits any discussion of Taylor's lack of leadership or Baikie's need to assume control of the ship. He is lavish in his praise of all the remaining members of the crew, singling out May and Crowther for special commendation. He goes into detail regarding the success of the specimen collecting, but states he believed "the trading part of the mission has been a great failure, but from causes beyond my control."[125]

Why he wrote this is unclear. Although not enough was made to cover Laird's costs in building and outfitting the *Pleiad*, the mission did turn a profit that far exceeded the cost of goods given in trade. His comment is perhaps a reflection on the lack success of the traders left at the confluence and their inability to trade due to their illness. But those who have written about this voyage have universally taken exception to it being called a "great commercial failure." Ajayi states that Macgregor Laird "made a substantial profit."[126] In fact, Dike says Laird's faith and vision in backing this mission "bridged the decade of disillusionment" in the Niger as an avenue for trade. He further states that the voyage of 1854 marked the end of exploration and that the traders took over where the explorers left off.[127]

On January 6, Baikie was finally at the rail of the *Bacchante,* which was prepared to steam out of Sierra Leone's harbor. As he waved goodbye to Heddle and William, he must have wondered if anything else would delay the return home. But luck was finally with him and good weather allowed the rest of the voyage to pass quickly. On February 3, 1855 they reached Plymouth Sound.

THE TOAST OF LONDON

With the ship tied up at the dock, Baikie said goodbye to May and began to supervise the offloading of his goods. His personal luggage amounted to very little. Clothing and private items were stored in a small trunk and the nets and other collecting equipment that had taken a half day to load on the *Forerunner* for the outward voyage had been left on the *Pleiad* to be transported back with the crew. However, in its place were now five wooden crates holding the over 900 items that Baikie had collected during the voyage. He had been unwilling to leave them on the *Pleiad,* wanting to begin the work of finalizing his records as soon upon his return as possible.

The containers were soon offloaded. Baikie would be taking one of the cases with him. Most of the collection would be left at Haslar, the specimens to be sorted and catalogued when he returned from London. A driver and wagon were secured for traveling to the hospital. There, the cases were put into storage and Baikie retired to his quarters for what would be a very short night.

The following morning Baikie rose early. He traveled to the train station where he supervised the loading of the case he would be taking with him to London and boarded the train. The Great Western Railway had completed the wide-track line from London to Plymouth only a few years prior to his sailing. The crates were transferred from the wagon to the train, the driver paid and Baikie boarded for his journey to London.

The Geographical Society had made plans for Baikie's stay in London and they had made it as convenient as possible. He walked out of the station and onto the lawn that separated Paddington from the rear entrance of the Great Western Hotel. The hotel was also new, its construction having been completed to coincide with the final work on the station.[128] At the front desk Baikie would find that the Geographical Society had booked an open-ended reservation in his name. He signed in and collected the messages that would dictate his schedule over the next few weeks. Reports, meetings, and presentations would occupy much of his immediate future. Although the train journey represented the finest in luxury travel for the period, it was still a long and tiring experience. At the end of a long day, all he wanted was a meal and a rest.

The next few weeks proved to be even more hectic than one might imagine. Sunday, the day after Baikie's arrival was spent going over the charts and his journal in preparation for the various meetings on his schedule. The following day he set off for the first of what would be many visits to the Geographical Society of London. There, over breakfast in the Society's rooms on Whitehall Place, he made his initial report to Sir Roderick Murchison and George Egerton, the Second Earl of Ellesmere, who had followed Murchison as President of the Society.

Both men were greatly impressed with the summary that Baikie had produced and the Earl of Ellesmere advised his secretary to set a date within the following six weeks on which Baikie would address the members of the Society with a general summary of the voyage, specifics relating to the charting of the Benue River, and an overview of the newly discovered flora and fauna. The Society would also set additional lecture dates for its members and sponsor two other dates for Baikie to provide the public with an overview of all he had seen. The presentation to the Society members would come first. This would give Baikie limited time to assess his collections at Haslar Hospital, uncrate the entire lot, catalogue them, separate the specimens into four lots (Geographical Society, Kew Gardens, British Museum, and his personal collection), and prepare a presentation for perhaps one of the most learned societies in the world. However, Baikie immediately agreed to the schedule. At the end of their meeting, the Earl of Ellesmere told Baikie that he and May were being jointly nominated for one of the Geographical Society's two gold medals. The final selection would take place in a few weeks and the presentation would be in early June.

The following day he met Sir William Jackson Hooker, the director at Kew Gardens. Hooker had been the head of Kew since 1841 and was aware of Baikie's earlier botanical publications. Having made own his first voyage as a naturalist and explorer to Iceland in 1815, Hooker also had more than a casual interest in Baikie's endeavors. For hours the two explorers, one in his seventies and one not yet thirty, traded stories and shared common experiences. On this, and on subsequent visits, they were joined by Joseph Dalton Hooker, who would soon succeed his father as head of Kew Gardens, and by George Bentham. Now retired, Bentham had recently donated his herbarium, numbering more than a hundred thousand specimens, to Kew Gardens. While Hooker was perhaps more interested in the stories of the collecting, Bentham was fascinated by the collections themselves and spent hours poring over the few examples that Baikie had brought to the meeting. Baikie was told that the genus *Baikiaea* was to be named in his honor, (Figure 8.1, *see colour section*) which must have added to his feeling of success. Hooker his son, and Bentham were among the most noted botanists of their time, and Baikie would have reveled at being in the presence of these three great men.

Macgregor Laird had traveled down from Liverpool to meet Baikie in person. They dined together at the Travelers Club. Located at 106 Pall Mall, the original concept of the club was to offer hospitality to distinguished English travelers and foreign visitors. The original rules excluded from membership anyone "who has not traveled out of the British islands to a distance of at least five hundred miles from London in a direct line."[129] Baikie admired the club's impeccable service, the dark oak panels, the polished silver of the dining room, the magnificent library, and the lively conversations over billiards. He must have felt like the most important man in London. Amid all the lords, cabinet members, and members of parliament it was Baikie who drew all the attention. It seemed as if everyone wanted to be introduced to him, to shake his hand and to enquire about his future.

Laird was happy with all commercial aspects of the voyage and was elated with Baikie's accomplishments. A network of trading stations from the delta to the confluence and beyond had always been part of Laird's dream. He asked if Baikie thought this was now possible. Baikie assured Laird that an abundance and diversity of trade goods were readily available and that the numerous people he met were ready to trade. Laird asked if Baikie would consider resigning his commission and heading the entire operation from an office on Fernando Po, or on the Niger itself? But Baikie would not consider move so serious as leaving the Royal Navy without first discussing it with Sir John Richardson, his trusted counsellor. Laird said that he understood, and the offer would remain open while Baikie considered his options and talked to his colleagues and superiors at Haslar.

An additional surprise event of those early heady days was an invitation to meet Lord Clarendon, Secretary of State for Foreign Affairs. Although currently deeply involved with maintaining the alliance with France, as they jointly battled against the armies of Tsar Nicholas in the Crimea, Lord Clarendon astounded Baikie with his general understanding of the Niger region and his specific knowledge of the *Pleiad's* voyage. They spent a pleasant hour together, which ended with Lord Clarendon asking about Baikie's interest in a West African consular post, should such a position become available.

As exciting, stimulating, and flattering as these conversations were, Baikie was anxious to return to Haslar. He needed a familiar and comfortable space to work with his collections and to put them into order in preparation for his series of lectures and presentations. He also needed time to discuss the offers that had been laid before him during the past week with Richardson. After his lengthy absence, Baikie took some time to visit his family and friends in Kirkwall. He then returned to London, and after a few brief days, he boarded a train to Gosport and his return to Haslar Hospital.

CHAPTER NINE

RETURN TO HASLAR

I n the fields outside London the stands of birch trees, their trunks surrounded by water from the heavy winter rains, stood scattered in the lower areas of large open pastures. The end of the rainy season had forced Baikie off the Niger. His return to England seemed to be marked by the wettest winter season he could remember. As the train moved south toward the coast, the skies began to clear, and the passing farms and villages were soon bathed in the vibrant sunshine of the English winter. Soon the flat fields and birch trees were replaced by rolling elevations with large stands of pines. As quickly as the hills had emerged, they vanished and soon the wrens and robins had been replaced by sea gulls. Portsmouth harbor and Haslar Hospital lay just ahead. Baikie looked forward to discussing the details of his voyage with Richardson and his other explorer colleagues.

THE MEN OF HASLAR

For over 250 years the Royal Hospital Haslar had had a long and distinguished history in providing medical care to the Royal Navy and later to all branches of service. The site of the Royal Hospital was formerly Haslar Farm and planning for the hospital commenced in 1745. The construction took sixteen years and was completed in 1762. A long line of prominent physicians had worked at Haslar, beginning with James Lind, "the Father of Nautical Medicine" who served as physician-in-charge at Haslar for twenty-five years. For most of the nineteenth century the Haslar Hospital produced a disproportional number of explorers and naturalists. This was due to a large extent to the posting at Haslar Hospital of two of the most famous travelers of their time: Sir John Richardson and Sir Edward Parry.[130]

At the start of the nineteenth century there were two ways for an officer to earn promotion in the Royal Navy. The first involved distinguishing oneself in battle. The second was to earn a reputation as an explorer and naturalist. Most of those serving as naturalists on expeditions were medical doctors. A single assistant surgeon could provide medical support. But, with their advanced scientific training, these same surgeons could also serve as the scientific officers in charge of assembling, cataloguing, and transporting the specimens collected.

When Baikie was first assigned to Haslar he worked under the surgical direction of Richardson while the hospital administration was under the command of Parry. In addition to their work at Haslar, Parry and Richardson had similar professional experiences in common. Both were lifelong military men who had made their reputations, not in battle, but as explorers and naturalists. Both had earned their reputation exploring northern Canada and the Arctic, and both had been knighted in recognition of their efforts. Although Baikie worked much more closely with Richardson, he saw Parry's guidance as equally valuable.

When Baikie was selected for the Niger expedition it had been Parry who told him about Beecroft and how fortunate Baikie would be to serve under him. Parry had talked of Beecroft's service in the Napoleonic Wars and how he had been captured by the French and held for nearly ten years. He spoke of Beecroft serving under him on his second Arctic expedition as they explored the coast of Greenland. It was this added knowledge of Beecroft's strength and leadership that had encouraged Baikie to serve on the Niger expedition and it was this knowledge that had led him to feel so inadequate when Beecroft had died before the voyage could begin.

Parry, who was born at Bath in 1790, was considerably older than Baikie. The son of a doctor, Parry joined the Royal Navy in 1803 at the age of thirteen and had made his mark long before coming to Haslar. In fact, his posting to Haslar was to mark the end of a very successful career. As one of the officers assigned to the brig *Alexander,* he made his first voyage to the Arctic in 1818. This expedition, under the direction of Sir John Ross, was in search of the Northwest Passage. The following year he obtained the chief command of a new Arctic expedition, returning to England in 1820 after a voyage of unprecedented success. Upon his return he was promoted to the rank of commander.

Parry had then been put in command of three subsequent Arctic expeditions and was credited with numerous discoveries. The third voyage in 1824–1825 was followed by an attempt to reach the North Pole by sled. Although it was not entirely successful, they reached 82 degrees 45 north latitude, which remained for nearly fifty years the highest latitude reached by any explorer. Parry was knighted in 1829. In 1846, after his retirement from active service, Parry was appointed Captain-Superintendent at Royal Hospital Haslar, and he was in that position when Baikie was first posted there. As Captain Superintendent he served as the overall officer in charge, controlling all hospital operations not directly related to medical care and he served in this capacity until 1852.

Sir John Richardson, upon Baikie's arrival, was serving as medical inspector and had responsibility for all aspects of staffing and conditions within the hospital itself. Richardson was, like Baikie, a Scot and was born at Dumfries in November 1787. At twelve he was apprenticed to his surgeon uncle. By fourteen he was enrolled in the medical school that Baikie had attended in Edinburgh University. By eighteen he was house surgeon to the Dumfries Infirmary. Like Baikie and Parry, Richardson had joined the Royal Navy. But in addition to his later exploits as an explorer and naturalist, he had first proven himself in battle, earning numerous commendations for bravery. He had seen action at the bombardment of Copenhagen, the blockade of the Russian fleet in the Targus and the attack on Washington in the War of 1812. Following the end of the war of 1812, he returned to Edinburgh where he gained his medical doctorate.

Despite Richardson's distinguished war record, it was his work as a naturalist that ultimately brought him recognition in the Royal Navy. His fame as an explorer and naturalist began with Sir John Franklin's voyage of 1819–1822. Richardson served as surgeon and naturalist and was responsible for documenting the geology, botany, and ichthyology for the official record of the three years spent in the Canadian Arctic.

Richardson published the narrative of Franklin's voyage in 1823, followed by a joint publication with Parry describing the species of birds and mammals collected by Captain Parry during his 1818–1820 voyage. In late 1823, Richardson again traveled with John Franklin. On this second Arctic expedition they surveyed the Mackenzie and Coppermine rivers in upper Canada. Upon his return, Richardson published his four volume *Fauna Boreali-Americana* detailing the numerous specimens collected on his most recent voyage.

In 1838 Richardson arrived at Haslar as chief physician. He would remain in the position for the next seventeen years. After forty-eight years of service, including three Arctic expeditions, he was not content to ease into retirement. He continued to complete numerous publications for another ten years until his death in 1865.[131] Although Richardson left Haslar Hospital in 1855, he would remain Baikie's mentor, confidant, and advisor for the remainder of Baikie's life.

Once in their positions of command at Haslar Hospital, Parry and Richardson surrounded themselves with doctors who shared their interests as naturalists and explorers. Baikie counted among his friends and associates at Haslar a number of doctors who owed their reputations to honors won through serving as naturalists on voyages of exploration. These men shared a common bond through their postings at Haslar Hospital and the opportunities, fame, and experiences provided to each of them by their association with Parry and Richardson. Baikie's fellow surgeons included Alexander Armstrong, Andrew Clark, Thomas Henry Huxley, and John Watt-Reid. Along with Richardson and Parry these individuals had created the strongest assemblage of explorers and naturalists in England, and perhaps in the world. It was to this auspicious company that Baikie had been posted in 1851.

Sir Alexander Armstrong was born in County Donegal in 1818. He studied medicine in Dublin and at the University of Edinburgh, and he joined the Royal Navy as assistant surgeon in 1842. Armstrong was commended by his superiors for effecting improvement in naval hygiene and in 1849 was promoted to surgeon. That December he was appointed surgeon and naturalist on the *Investigator*. The ship was one of two vessels commissioned to sail to the western Arctic by way of the Bering Strait in search of the ships of Sir John Franklin, missing since 1845. They left England in January 1850 and finally reached Mercy Bay on the northern shore of Banks Land, where the *Investigator* spent the winters of 1851–1852 and 1852–1853 beset in ice.

The *Investigator* was not able to break free from the ice and the captain ordered the ship to be abandoned. Armstrong's plant and animal collection perished with the *Investigator*, but he managed to keep his journal. Upon his return to England he was posted to Haslar. His personal narrative of his time on the *Investigator* was awarded the Gilbert Blane gold medal for the best journal kept by a surgeon of the Royal Navy. In 1858, while at Haslar, Armstrong's observations on naval hygiene and scurvy were published in London. He spent the following decades on seagoing appointments in the Baltic Sea and the West Indies, as superintendent of the naval hospital at Malta, and as director-general of the Royal Navy's medical department. He was knighted in 1871 and in 1880 retired to a quiet life. He died in 1899.

Another Scot, Andrew Clark, was born on October 28, 1826 in Aberdeen. His mother died during his birth and his father, a doctor, died when Clark was seven years old. Two bachelor uncles directed his education and he studied medicine at Edinburgh University from 1843–1846. Baikie and Clark were the same age and had known each other for years. They studied medicine together at Edinburgh University and Clark and Baikie were posted to Haslar at the same time.

Like Baikie, Clark had joined the Royal Navy and held a commission from 1846–1853 as an assistant surgeon. Clark spent most of those seven years at Royal Haslar Hospital, where he specialized in pathological work and pioneered employing the type of microscope usage that was to become the recognized practice for the profession. As with other Haslar doctors, Richardson's influence secured Clark's services as naturalist and assistant surgeon on an expedition that took him on an exploring and collecting voyage to Madeira. When he left the Royal Navy, Clark was appointed curator of the museum at the London Hospital, reflecting his enthusiasm and expertise as a naturalist.

In 1854 Clark became a member of the Royal College of Physicians. He held a lectureship at the London Hospital and was elected a fellow of the Royal College of Surgeons. During the cholera epidemic of 1866 he became friends with, and physician to, William Gladstone. Clark was elected a fellow of the Royal Society and president of the Royal College of Physicians. In 1883 he was created a baronet in recognition of his services to medical science. Clark suffered a stroke in 1893 and died two weeks later.[132] During his illness Queen Victoria desired that she be kept informed daily of his condition.

Also at Haslar Hospital at the time of Baikie's arrival was Thomas Henry Huxley, who was about Baikie's age, having been born in London on May 4, 1825. In 1842 he entered the medical school attached to Charing Cross Hospital as an apprentice to his uncle. Huxley passed the M.B. examination at the University of London. In 1846, after briefly practicing medicine in London, he entered the Royal Navy. Huxley was posted to Haslar Hospital, but soon, through his association and friendship with Richardson, he was selected to be assistant surgeon aboard the *H.M.S. Rattlesnake*. This voyage was assigned to survey and chart the territory around Australia and New Guinea. During the voyage Huxley collected and studied marine invertebrates, sending his papers and specimens back to London. Upon his return Huxley was elected a fellow of the Royal Society. However, his numerous attempts to get the narrative of his discoveries published by the government were not successful and he left the Royal Navy in 1853. In 1856 he met Charles Darwin and became a proponent of Darwin's theory of evolution by natural selection. Huxley's repeated and passionate defense earned him the nickname of "Darwin's Bulldog." Huxley continued to lecture, gather data, and publish, mostly about human origins. From 1881 to 1885 he was president of the Royal Society and in 1892 he was appointed to the Privy Council. Huxley died on June 29, 1895.

Finally, there was John Watt-Reid, also a classmate of Baikie's at Edinburgh University. Sir John Watt-Reid was born on May 20, 1823 in Edinburgh. His father, a doctor, provided for his education at Edinburgh Academy, Edinburgh University, and the extra-mural medical schools. He entered the Royal Navy in 1845 as assistant-surgeon. He was initially posted to the Royal Naval Hospital Plymouth, where he received a commendation from the Board of Admiralty for services during the cholera epidemic in 1849. As a result, he was promoted to surgeon and posted to Haslar in 1851. Watt-Reid was assigned to be commander-in-chief in the Black Sea Squadron on the flagship *Britannia*, where he again received a commendation for services to the sick during another cholera outbreak.

Watt-Reid was next ordered to the *Beileide* hospital ship serving in the China War of 1857–1859, and in 1866 he was promoted to staff surgeon. In 1874, for his work during the Ashanti Campaign in West Africa, he was promoted to deputy-inspector-general. He went on to become inspector-general and finally medical director general. He was made honorary physician to the Queen in 1881 and was knighted in 1882.[133]

This cohort of four brilliant young surgeons, along with Parry and Richardson, would collectively publish over 100 books on all aspects of natural history. All were Royal Navy surgeons and most had studied medicine at the University of Edinburgh. Individually, they explored Canada, the Arctic, the West Indies, Madeira, China, Australia, New Guinea, and West Africa. As doctors, they improved naval hygiene, perfected the treatment and prevention of scurvy, cholera, and malaria, and piloted the use of the microscope in pathology. Five would be knighted before the end of their careers. Two would become directors general of the Royal Navy's Medical Department. One would become inspector-general of the Royal Navy. One would be named president of the Royal College of Physicians, and one president of the Royal Society. Three would become honorary physicians to the Queen. Despite all the dignitaries whom he had met in London, it was with this group that Baikie was most anxious to discuss his recent voyage.

THE LOSS OF RICHARDSON

Baikie supervised the offloading of his remaining crate and hired a cart to take it to Haslar. He could have also ridden on the cart but instead chose to take the short walk from the train. As he moved toward the main gate, he could see beyond to the central arch marking the main entrance to the hospital. Once inside, Baikie turned to his left toward Richardson's house, but no one seemed to have noticed his arrival. About half-way between the central arch and the chief physician's residence, Baikie could see the entrance to the cellar. Richardson had established the practice of utilizing one of the large, vaulted cellar sections as an area to sort and catalogue the specimens he had collected on his various explorations (Figure 9.1). Richardson offered the use of these areas to other members of the Haslar staff whom he had recommended for earlier voyages. Thus, before Baikie had left on the Niger Expedition, Richardson had offered him the use of the cellar for a storage and work area.

With the assistance of one of the orderlies, it was in this designated location that Baikie now placed his materials from London with the rest of his crated collection. Baikie was

Figure 9.1 Richardson's museum at Haslar © Haslar Heritage Group

tempted to begin the unpacking immediately, but instead secured the door and moved back outside into the fading sunlight. He walked to the end of the hospital and headed for the junior officers' mess on the first floor. There, assembled and awaiting his arrival, were the officers from both the post and hospital. At the head of the table, now rising to propose a toast, was Sir John Richardson.

Although Parry was gone, Richardson still remained at Haslar and Baikie now joined him at the head of the table and began to talk of the voyage, his collections, and of his reception upon his return to London. Clark and Huxley were also gone, but those who remained and the newcomers to the group, seemed delighted at his success and interested in all that he had seen and done. As the food arrived the talk around the table diminished and Baikie asked for Richardson's guidance and support as he prepared for his upcoming presentations and the creation of his narrative of the voyage. Richardson then shared with Baikie what others around the table already knew.

Richardson had been a candidate for the post of director general of the Royal Naval Medical Service. The announcement of the nomination had been made before Baikie had left for the Niger and it was assumed that the appointment would have been made while Baikie was on the river. This would have limited Richardson's time at Haslar; however, he would still have been able to provide the input and guidance that Baikie needed. Richardson relayed that within the past week it had been announced that he had been passed over for the post. Richardson had announced his resignation and he would be leaving both Haslar and the Royal Navy within a few days.

The euphoria that he had felt on walking into the room now dropped away. At twenty-nine years of age Baikie was at the peak of his profession. He was meeting heads of state and the top scientists of his day. Yet this sudden loss of his esteemed instructor and advisor would have taken all the joy and celebration from what should have been his triumphant return. What Baikie could not know was that this event would mark the turning point for all that would follow for the rest of his short life.

In the second week following his return to Haslar, Baikie had two of the now sorted cases sent to the British Museum. He then traveled to London to meet Sir Augustus Wollaston Franks, the recently appointed head of the museum, where the two pored over the collected items. With his own emphasis on ethnography, Franks too was interested in the artifacts and the stories of their collection. This meeting, like those previously, was successful and encouraging and helped to ease the feelings of loss that Baikie felt with Richardson's departure from Haslar.

Following this meeting, Baikie returned to Haslar; spending much of this time preparing his presentation for the Geographical Society. This was possible because Haslar Hospital did not function in the same way as the civilian hospitals in the area. Most of the Haslar patients had been wounded in action in the Crimea. In these circumstances a crewman onboard a Royal Navy ship who was wounded in battle would have been treated by naval surgeons on board the ship. Those with minor wounds were tended to and, when they recovered, returned to duty. Those with major injuries requiring significant surgery, such as an amputation, would have the procedure done on board the ship and then would have been sent home. These sailors were then admitted to Haslar, where they would complete their treatment and recovery before returning to civilian life.[134]

Instances of major surgery at the hospital were rare. Occasionally an injury would occur close enough to the hospital for the wounded man to be brought directly to Haslar. Sometimes an operation conducted onboard ship would have gone wrong, requiring the surgeon at Haslar Hospital to reopen the wound and repeat the procedure. But generally, Baikie's responsibilities required his presence in the hospital only twice each day as he made his rounds. He would check in on each of those assigned to his care and make recommendations for continued treatment, or recommend to the chief surgeon that a patient was well enough for release.

This left Baikie a great deal of time to examine and catalogue his collections and to prepare for his upcoming presentations. Baikie would rise, dress, and go directly on his morning rounds. Following the rounds, he would take breakfast and then go to the cellar where his collections were now spread around most of two huge vaulted rooms. He would methodically systematize and sort items into groups to be shipped to the appropriate museums, setting aside key items that he would use to illustrate each of his upcoming presentations. Lunch would be followed by his afternoon rounds. Following those duties Baikie would retire to his room above the officers' mess to work on the narrative of his voyage until dinner. After the evening meal he would return to the vaults and continue working on his collections until late into the night.

The work on the collections was going well. He was now comfortably prepared to address each of his varied audiences on the designated topics. However, the writing was not going as well. This was his first attempt at narration. His other efforts had been natural histories and he found it difficult to put his own accomplishments into words. The expectations of his upcoming lectures and presentations were very high—higher than had ever been expected of him previously. He would have valued input from Richardson as to how to traverse the political and scientific waters into which he now found himself immersed. However, Richardson was gone and Baikie missed his friendship and guidance.

In those early days Baikie must have felt alone. He was the only one among the assembled naturalists at Haslar who had yet to make his mark. Now he was back with his own stories to tell but there was no one to listen. Most of his companions had left both Haslar and the Royal Navy. There were close friends with whom he worked, perhaps there were those in whom he could confide. But it was not the same. The hospital that once housed some of the best and brightest naturalists and explorers was now a different place. Baikie faced the loneliness that one encounters when there is no one who has shared the same experiences. He became more withdrawn, spending all his time preparing for his presentations and keeping his own counsel.

PRESENTATIONS AND DISAPPOINTMENT

In everyone's opinion the initial presentation to the Society had gone well. However, Baikie himself had not been as pleased. Presentations on the plants, animals, and artifacts collected had all appealed to the narrowly focused audience at the Geographical Society of London. But Baikie had hoped his successful use of quinine as a preventative to fever would excite everyone and he had made this point a key component in his presentation to the Society. To his dismay and disbelief, proving that the use of quinine could remove the greatest barrier to the white man exploring and trading in Africa seemed to be lost on both the listeners and newspaper reporters covering the event. His meetings and presentations also seemed to be falling more and more under the constant shadow of Dr. Heinrich Barth.

Barth, the man whom Baikie had been sent to rescue, had in fact passed through the region Baikie had just explored. However, as he had later discovered, Barth was already nearing the coast as Baikie was headed upriver. Furthermore, by traveling alone Barth had not experienced the delays that Baikie had encountered. Thus, Barth had not only not been rescued but had actually beaten his potential rescuer back to England by several months.

Many of the groups to whom Baikie was scheduled to speak would have heard Barth's presentations only a few weeks or even days previously. The areas of exploration and topics were the same for both travelers. Baikie had traveled by boat. Barth had done much of his exploring on foot. Baikie had been in Africa for months. Barth had been in Africa for years. Taking away the medical discovery of quinine that seemed to capture no one's interest, Baikie believed he would be received by the audiences as a poor second best.

After his initial presentation to the Geographical Society's members, neither the additional member programs nor the public appearances were forthcoming. Baikie assumed the Society was busy with the nomination and selection process for the Society's gold medals. In early June, he once again traveled to London for the award presentation. Baikie assumed that the prestigious patron's medal would go to Barth. It was fair. There was no comparison in length of time or distance covered. However, Baikie assumed he would be the recipient of the founder's medal.

Both men had been sponsored by the Geographical Society of London. The medals were said to be of equal status and each was awarded "for the encouragement and promotion of geographical science and discovery." But, even if it was considered a second place, the recognition provided by the founder's medal would be a fitting tribute to an enterprise that everyone was calling a great success. Following his earlier lecture for Society members, the Earl of Ellesmere had reminded Baikie of the date of the award presentation and he clearly had wanted to make certain that Baikie would be present. Baikie assured him that he would attend and assumed this to be yet another indicator that he was to be one of the recipients. As it turned out, Baikie was wrong on all counts.

On the evening of the event Lord Ellesmere, in his role as Society President, outlined the work that they had sponsored during the previous year. Explorers who were present, including Baikie, were asked to stand and were introduced to the members. Then the recipient of the patron's medal was announced. "David Livingstone: for his recent explorations in Africa." The announcement was met by a standing ovation and Baikie joined in the hearty recognition. Livingstone gave a brief acceptance speech and the Society's President again took the floor to award the founder's medal and the prepared statement was read. "To Charles John Andersson: for his travels in South Western Africa."[135]

What Baikie could not have known was that 1855 had been a banner year for nominations for the Geographical Society's two gold medals. This was true both in numbers and in the quality of the nominees. In a typical year the Society would get four to six nominations to be considered for the two awards. When the Society had met on March 26 to decide on the medal winners for 1855, Baikie had been one of twelve individuals nominated. Livingstone had just made a transcontinental journey across Africa from the Atlantic to the Indian Ocean and had done it on foot. Andersson, the Swedish explorer, had trekked through the interior of southern Africa since 1850. Upon his recent return to London in 1855, he had published his book, *Lake Ngami,* which had been an instant success.

Livingstone and Andersson had been selected. But the nominees who did not receive either medal included some of the most noted names in science and exploration. In addition to Baikie and May, Heinrich Barth, Eduard Vogel, Richard Burton, Joseph Hooker, and Thomas Thomson were not selected. Other nominees included Evariste Huc, a French Catholic missionary and explorer who studied the Tartars, Mongols and Tibetans and had recently published *Travels in Tartary, Thibet and China;* Mansfield Parkyns, who had extensively explored the area around what is now Ethiopia and Eritrea, and had also recently published *Life in Abyssinia;* Sir Henry Young, explorer and governor of South Australia; and Dr. Hinrich Rink, the Danish geologist, explorer and foremost expert on Greenland.

But Baikie knew none of this and must have been stunned and disappointed. He had been asked to attend and had assumed that this meant he would receive one of the awards. Instead, neither he nor Barth had been selected. As the crowd began moving toward the reception, Lord Ellesmere made a point of intercepting both Baikie and Barth. Baikie was told that the timing of his return, fewer than two months ahead of the meeting to select this year's recipients, had hurt him. Neither he nor Barth had completed the narrative of their travels, and in Baikie's case he had made only one presentation to a limited audience. They were reminded that nominees were often carried forward to subsequent years and Lord Ellesmere was confident that they would both be selected, perhaps as early as the following year.

Baikie did not stay for the reception but instead caught a cab back to the Great Western Hotel. He quickly packed and caught the night train back to Plymouth. It was a long, dark ride with ample time to think. Did Baikie try to convince himself that he had been foolish? The award was not relevant, he had been told the gold medal would come in the future. He had been successful. Everyone had said so. His course was simple and straight forward. He would finish the narration of the first voyage, get it published, and then tell Macgregor Laird that he was ready to go back to Africa.

CHAPTER TEN

THE NIGER REVISITED

B aikie's triumphant return in 1855 was followed by a series of events that set the stage for the devastating last decade of his life. His return to the Royal Hospital Haslar had evolved from being an eagerly anticipated homecoming to a monotonous routine that he must have looked forward to less and less each day. With Richardson's departure and the dispersal of his explorer colleagues, there were only his medical duties to keep him occupied. His earlier presentation to the Geographical Society of London had been his only appearance before this auspicious group, and his scheduled public presentations never materialized at all. His collections had been divided and distributed to the various museums that had expressed an interest in them, with most of the items going to Kew Gardens and the British Museum, but he was now alone.

GEOGRAPHICAL SOCIETY OF LONDON

Like all nominees for the Society's gold medals, Baikie had been provided a copy of his and May's nomination as recorded in the Society's meeting minutes from March 26:

> For having volunteered on the death of Captain Beecroft, to undertake the hazardous ascent of the Kwarra and Chadda in the Steamer Pleiad which they accomplished to a distance of 250 miles on the Chadda River, beyond the farthest point reached by the previous expeditions of Allen and Oldfield and remaining on the River for months without any loss of life, also for correcting previous surveys of the Kwarra by a careful series of observations and extending the survey of the Chadda to the limits of their exploration, thereby affording data for estimating the correctness of the unestablished positions of recent travelers between the Chadda and Kwarra. Also for investigations concerning the

political, commercial and natural geography of the regions
visited, an outline of which has already been communicated to
the Society, together with a map. It is also to be noticed that one
of the results of the judicious and conciliatory conduct of this
enterprise is the opening of a road into equatorial Africa, from
the West, thereby obviating the obstructions in the way of that
object which have attended the attempts of recent travelers from
other quarters.[136]

Although it was flattering to read, this must have caused further confusion for Baikie. Why
had he and May been jointly nominated? True, both had worked on the mapping, surveying,
and charting of the Niger and Benue rivers. However, May had been a last-minute volunteer
who was not part of the original delegation. Baikie had been commended for volunteering to
head the mission, for his investigations of the political, commercial, and natural geography of
the regions, and for his preventive use of quinine which resulted in all members of the party
returning safely. May had played no part in any of these efforts.

In addition, there was no history of the Geographical Society putting pairs of individuals
forward for consideration. In fact, it was quite the opposite. Richard Lander had traveled
with his brother John when tracing the route of the Niger River. Richard Lander was the
initial gold medal winner. His brother was not included. Livingstone was the most recent
winner. John Kirk had been with Livingstone for his entire time in Africa. Unlike May, he was
not mentioned as a potential recipient with Livingstone. Hooker and Thomson had served
as joint naturalists, as had Vogel and Barth, on their two expeditions. However, they were
each nominated independently, not as part of a team. Curiously, in the entire history of the
Society's gold medal awards, there has been only one pair of explorers who were jointly given
the award and that was to a husband-and-wife team. This award was not made until 1991.

Because the Geographical Society of London had been one of the voyage's sponsors, Baikie's
first task had been to write an extensive description for publication in their journal. This had
included the maps May had drawn from Baikie's original surveys and charts explaining the
Pleiad's ascent of the Niger and Benue rivers. His intention was to continue expanding this
writing into what would become the published narrative of his voyage. The initial article had
been written, submitted, and printed in the Society's journal and had been well received. The
exception had been voiced by Rear-Admiral Fredrick Beechey, who had recently been named
president of the Geographical Society. Beechey had referred to the article as "interesting and
instructive," reflecting "credit upon its author," but then went on to say:

I must not omit to notice an oversight which I am sure Dr. Baikie
will, with his usual candor, acknowledge. In alluding to the origin
of the expedition, Dr. Baikie does not mention the persevering
part taken by the Council of this society, and particularly by Sir
Roderick Murchison, in promoting it; and he entirely omitted to

connect the name of McLeod with the great and novel feature of the
plan which rendered this expedition so successful in all respects,
and will govern the operations, in regard to season, of all future
expeditions. It will be seen in our Journal that, early in 1852, a project
for ascending the Niger *with the rising waters*, was laid before the
Council by Lieutenant MacLeod, who had been employed for some
time on the Africa coast. Having been referred to the Expedition
Committee, attention was directed to a clause in Mr. Laird's mail
contract with the Admiralty, which provided for the ascent of one of
the African rivers by steam, at a small expense; and the Committee
recommended Lieutenant McLeod to communicate with Mr. Laird
and adopt his plan to this arrangement. Other steps were also taken
and communicated to the Society by Sir Roderick Murchison, in
his Presidential Address of that year. In 1853 the expedition having
been brought under the notice of the Government by Sir Roderick,
as President of the Society, some progress was made. In 1854 the
expedition started, and it was intended that the veteran African
explorer, our late member, Mr. Consul Beecroft, should take the
command; but his lamented decease having occurred a few days
before the arrival from England, the command devolved upon
Dr. Baikie. I have felt it to be due to the persevering efforts of this
Society in promoting this expedition, and to the individuals whose
names are so honorably connected with it, to insert in some detail
these facts connected with its origin; of which, I am sure, Dr. Baikie
will acknowledge the justice and propriety.[137]

This was a stinging criticism. His writing was intended as a summary of his voyage that
would be followed by a much more comprehensive manuscript outlining the full details of
his journey. Of course, the Geographical Society of London deserved credit for the venture,
and Baikie thought this initial piece had done just that. The article had been published in
their journal, and only the efforts that had been supported by the Society were ever featured
in their publication. However, the slight outlined by Beechey had extended well beyond the
perceived snub of the Society. Beechey had publicly accused Baikie of taking credit for a plan
that had been conceived by MacLeod and for publishing maps that were the work of May. As
to his leadership of the expedition, it was clear that the Rear-Admiral reduced his role to that
of a place holder who had simply kept the expedition together upon Beecroft's death.

Nowhere was there any mention of his assumption of command and his moving the voyage
forward when Captain Taylor had wanted to return to the delta. Nowhere was there praise
for the commercial and scientific success of his efforts and nowhere was there a mention that
this had all been accomplished without loss of life due to his insistence upon the regimen of
quinine. For the second time in as many months Baikie's work had been disregarded by the
Society.

NARRATIVE OF BETRAYAL AND TREACHERY

Baikie had quickly created a draft account of the expedition. He combined this narrative with the maps, charts, and accounts that he and May had compiled and had submitted the completed proposal for publication. However, both Samuel Crowther and Thomas Hutchinson had already published narratives of the voyage. Baikie described both of their publications in the preface of his book. Crowther's is described as "from a most estimable individual, is in the form of a short journal, referring more especially to the prospects for missionary efforts." But Hutchinson's work is described much less positively as relating to the "private opinions and individual experiences of a member of the party, (rather) than to contain an account of the expedition."[138]

Hutchinson had selected a title almost identical to the one chosen by Baikie. Hutchinson wrote the *Narrative of the Niger, Tshadda and Binue Exploration.* Baikie wrote the *Narrative of an Exploring Voyage of the Rivers Kwora and Binue.* But Baikie's feelings toward Hutchinson were not a reflection on the title, or even that Hutchinson had been able to go to print several months ahead of his release. Instead, they derived from the fact that the Irishman had described the quinine experiment in detail in his book, without mentioning Baikie's role in the process.

Hutchinson lists himself as "senior surgeon" on board the *Pleiad.* This is accurate, although the appointment was made by Baikie after he had been promoted to leader of the expedition. Hutchinson goes on to state that, in this capacity, he "conducted research on the use of quinine as a preventative measure against the effects of malaria." He even describes the methodology and indicates that "in small doses, (taken with wine) quinine had a favorable effect in preventing fever." Claiming sole credit for the use of quinine as a prophylactic completely contradicts the version in Baikie's narrative. Baikie stated in his journal, written at the time the action was taken, that he had turned all the medically related issues over to Hutchinson except for the distribution and monitoring of the quinine dosage.

But Dr. Hutchinson continued to press his claim. At an event celebrating the Geographical Society of London's anniversary, the Earl of Ellesmere mentioned Baikie's pioneering use of quinine as one of the highlights of the previous year. Following this acknowledgment of Baikie's successful clinical trial, Hutchinson wrote the following letter to Rear-Admiral Beechey at the Geographical Society:

> Sir,
>
> On coming to the offices today I saw a copy of an address of the Earl of Ellesmere, read at the anniversary meeting of the Geographical Society, wherein his Lordship winds up his comments on the successful issue of the late Niger to Chadda exploration by observing this signal and encouraging success is mainly due to the skill and care of Dr. Baikie, R.N.

Allow me to inform you, and through you the respected President of the Society that with the sanitary results of our expedition Dr. Baikie had no more to do than any resident of Parliament. He was merely a passenger on board the *Pleiad* to conduct and superintend the scientific arrangements of the Admiralty part of our expedition and his "skills and care" had no concern with bringing the crew home scatheless. The hygienic and prophylactics means used to control fever were entirely my suggestion and under my control.

Thomas J. Hutchinson[139]

There is no indication that this letter, or the comments it contained, were ever shared with Baikie. Phrases like "Had no more to do than any resident of Parliament," "was merely a passenger on board the *Pleiad*," and "his skills and care had nothing to do with [the health of the crew] and the [treatment] was entirely my suggestion and under my control" take Hutchinson's claims to a much more personal level. However, even if the letter had been shared with Baikie it would not have been seen ahead of the publication of his book.

To offer some perspective on Thomas Hutchinson it is necessary to provide a glimpse of who he was before and after the voyage on the *Pleiad*. To his credit, he had traveled to West Africa in 1851. His knowledge of customs and the people would have made him an asset when visiting the local rulers. Baikie's accounts of visits with local rulers consistently list the participants and give credit for the role each played. Crowther, May, and Richardson were listed on every visit to the local villages. Hutchinson was identified by Baikie as having participated on only one of the visits. However, Hutchinson's book includes detailed descriptions of events and meetings occurring in many different villages where Baikie's journal does not include him among the participants. Hutchinson describes the same details as Baikie but fails to mention any of his colleagues by name. Reading Hutchinson's work, the reader would assume that he participated in every meeting and was traveling alone.

Hutchinson provided information for his biography, which appeared in the *Dictionary of Irish in Latin America*. He writes that following the expedition, in 1855, he was appointed consul for the Bight of Biafra at Fernando Po. He states that he held that position for two years. However, official British records indicate the consul was Benjamin Campbell from 1853 to 1859. Perhaps he was referring to the colonial governor at Fernando Po. Yet the official British records list the governor as John Beecroft from 1833–1854; and James Lynslager from 1854–1858. There is no record of Hutchinson ever holding a consular or governor's position in West Africa.

In 1858, Hutchinson was appointed consul in Rosario, Argentina, where he was also an agent for Lloyds. Thomas Murray, in his book about the Irish in Argentina, wrote that there were rumors in Buenos Aires that Hutchinson got his appointment and preference from the English government for "betraying his friends." "This could not be definitively proven, [but] Hutchinson's connections and friends were the principal cause of his appointments throughout his time in the consular service."[140]

Meanwhile, Baikie's book proposal had been accepted by his publisher and he had been given a small advance. This early payment, along with his naval salary which he had collected upon return to Haslar, had given him a modest amount of financial freedom. This allowed Baikie to leave the junior officers' quarters at Haslar and take up residence in a small house he had rented in Gosport. This move would have had less to do with his financial security and more to do with the lack of connection and camaraderie he felt towards the other surgeons with whom he now worked. Baikie now poured all his efforts into completing his narrative of the expedition.

The home in Gosport was where he now spent most of his time, devoting his days to writing and yearning for some change in his status that would allow him to once again feel the exhilaration and challenge he experienced while he was in command on the Niger. Working in the relative seclusion of his new home he completed the *Narrative of an Exploring Voyage up the Rivers Kwora and Benue* within a year and it went to press in 1856.

Baikie now laid claim to proving the prophylactic value of quinine. However, Hutchinson had taken the credit for himself much earlier in his own publication. Also, while Baikie relegated his findings to an appendix, Hutchinson had allocated a full chapter to malaria and "his" treatment. In his narrative Baikie writes:

> While up the Niger in 1854 I had ample opportunity of testing this virtue [of quinine] and I must unhesitatingly record my belief in its existence... I can affirm that after taking my morning dose (2–4 grams) I felt fit for any kind of duty, all the languor of a close damp tropical night dispelled, or in the evening after a hard day's work in the hot sun, nothing was so reviving and exhilarating as this invaluable drug.

In his account he claimed that "the great modern improvement is the discovery that quinine not only cures, but that it prevents, and that by taking this invaluable drug while in unhealthy localities, persons may escape totally unscathed." He also expressed his strong opposition to the medical community's current proposed treatment: "Drugs should be avoided as much as possible, especially calomel and other mercurials, which are not only unnecessary, but have actually killed far more people than the fever ever has." Neither the medical community nor the public seemed ready to grasp the importance of this writing, and the old methods of treatment continued for the next forty years.[141] Baikie would later develop his procedure into a long treatise for the British Museum.

Baikie's problems were not limited to his battle with Hutchinson, as Thomas Taylor reappeared. Captain Taylor filed charges of mutiny and piracy against Baikie for his having assumed control of a ship that had been under Taylor's command. In the preface of his book, Baikie states:

> I have felt obliged often to be more personal than I could have desired, and to allude to disputes and to differences over which I

would gladly have thrown a veil; but as these matters have lately been made the subject of judicial proceedings, I have mentioned them, partly in my own defense, partly to show why the success of the Expedition was not more complete than it was.[142]

Though he did not mention Taylor by name it is a clear reference to the legal problems he was forced to deal with at the time of his book's publication.

Between June 1855 and June 1856 Baikie was challenged on his leadership, his surveying and charting of hundreds of miles of the Benue River, and his medical breakthroughs on the use of quinine. He was effusive in his praise for Daniel May, who had volunteered to join the expedition the day before it left Fernando Po. Baikie was then accused of taking credit for the charting and surveying that May had done. He had voluntarily assumed control of the mission twice, once upon the death of Beecroft and once when Taylor had proven unable to see the mission through, only to be taken to court and accused of piracy and leading a munity. Finally, he was forced to prove that the quinine experiment had been his undertaking and not Hutchinson's, despite claims that had been published well ahead of Baikie's narrative. To add insult to injury, in the late 1850s, Bailey and Wills of Horsecley Fields produced "Dr. Hutchison's Quinine Wine," marketing it to ship owners and crews.

It is unclear whether May raised the issue of receiving inadequate credit with the Society, or if someone within the organization was serving as his advocate. If May was unhappy, it is apparent that Baikie was unaware of his feelings as he worked to make sure that May would travel with him on the next trip to the Niger. This was a decision both would come to regret. The court issue involving Taylor was dismissed with no action taken against Baikie. The case of who proved that quinine was a preventative to malaria became a nonissue. Although Hutchinson may have made some short-term gains through his medical sales, history failed to give either man credit for the work.

Where the impact of these multiple issues was felt was within the walls of the Geographical Society. Nominations for the 1856 awards were held April 14, 1856 and a much shorter list was developed. Heinrich Barth and Richard Burton were put forward again. Andrew Waugh, who explored and surveyed the Himalayan region; and Richard Collinson and Elisha Kane were each put forward for their efforts in exploring the Arctic region in search of the lost Franklin expedition. However, with Baikie's successes having been questioned or challenged on three different fronts, his name was not listed among those being considered. Nor would he be nominated any time in the future. The medals in 1856 were awarded to Heinrich Barth for his extensive explorations in Central Africa, his excursions about Lake Chad and his perilous journey to Timbuktu, and to Elisha Kane for services and discoveries in the Polar regions during the American expeditions in search of Sir John Franklin. In his acceptance speech, Barth heaped high praise on Baikie's efforts in attempting to locate him and to affect a rescue. However, Baikie did not hear these remarks. He was not in attendance.

LAIRD AND THE CMS

Though he would have been deeply hurt by the criticism of members of the Geographical Society of London, and while his experimental use of quinine had been generally ignored by both the public and medical profession, two of his accomplishments had been touted as major successes. The work he had done in his original capacity as naturalist had been roundly praised by the museum community. Also, his success at trade had made a rich man even richer and Macgregor Laird wanted to build on that success.

Until the voyage of the *Pleiad*, trade with the interior of Africa had not seemed possible. This Macgregor Laird-backed venture had changed the situation, as it had proven that Europeans, after taking preventative measures against malaria, could survive in the interior, which in turn allowed them to trade directly with the inland African people. The voyage had been a commercial success, but it had not offset the amount of money Laird had put into the building of the *Pleiad* and he could not afford to be the sole source of funding for the next, much larger effort.

The British government was being urged by business and religious interests to pursue the further exploration of the Niger River on a variety of fronts. However, British men and money were still flowing into the Crimean War. Laird and other merchants were pushing to tap the resources of the areas that Baikie's 1854 mission had opened. There was also an interest on the part of the Birkenhead shipbuilders in pushing farther into the interior on both the Niger and Benue to see how far up the rivers their specifically designed light draft ships could go. The CMS was lobbying for Reverend Crowther to return on yet another trip up the river to establish permanent church missions.

Without waiting for the promise of Laird's financial support, it was the CMS who decided to make the first move. This is a classic illustration of the fact that Christian missions were an agency of colonial penetration of Africa and in many instances, the Bible proceeded the flag. Soon after the Treaty of Paris ended the Crimean War, the CMS put a proposal to Lord Palmerston at the Colonial Office. Although England and France had been allies in the recent conflict, France was England's primary rival in African trade. Members of the Colonial Office were urging the establishment of a permanent British foothold in the region before France or one of the other European powers could move in to fill the void.

Faced with these compelling reasons for an additional mission, Lord Clarendon, Secretary of State for Foreign Affairs, agreed to a five-year contract with Laird to continue charting the Niger and its tributaries and to establish permanent trade stations at Aboh and Onitsha, and at the confluence of the Benue and Niger rivers. This meant Laird would receive eight thousand pounds during the first year, with the amount dropping at a rate of five hundred pounds per year for the next four years. This guarantee secured Laird's commitment to the venture.

Apart from trade and missionary considerations there was still a great deal of general exploration to complete. Nothing was more important than the accurate charting of the river channels and the examination of their course to determine how far upriver light draft ships could safely proceed. The Geographical Society of London was aware of this and gave their backing to the expedition as well. Lord Clarendon endorsed this joint proposal, and the project began to move forward.[143]

Like the previous mission of 1854, the Colonial Office was very specific in its stated objectives for the voyage. The primary mission would be to survey and chart the Niger River beyond the confluence, as they had done earlier for the Benue. And they would make treaties with the local leaders whom they had previously contacted to encourage trade and missionary work. This time they would establish the string of trading stations, churches, and schools. It was understood that, beyond the three sites specifically mentioned, Laird would be unrestricted in the location and structure of additional trading stations.

Also at this time, Reverend Crowther was given a definite commission from the CMS to establish the Niger Mission with its permanent church sites, as the organization's first totally African enterprise. Laird agreed to transport Crowther and his party and to assist them in their contacts. Through these various efforts, the government believed the group would be furthering British interests and influence, ahead of the other European powers.

The final objective was to be a reconnection with the Sokoto caliphate. Sokoto was by far the largest independent country in Africa. It stretched from what is now Burkina Faso to the Central African Republic and included the confluence of the Niger and Benue rivers. The British government had never confirmed the accord reached between Clapperton and Muhammad Bello. They wanted this to be concluded and Baikie to be accepted by Sokoto as the British government's representative.[144] The Foreign Office dictated the expedition should sail as far on the Niger River as possible. At that point, some of the group would then travel overland. They would deliver a letter written by Queen Victoria to the Caliph of Sokoto, explaining the reasons for the expedition, offering friendship, and asking for renewed discussions on the Clapperton–Bello accord, leading toward a permanent trading station and a long-term transactional agreement.[145]

THE EXPEDITION TEAM IS ASSEMBLED

It had been nine months since Baikie's book had been published and he was more than ready for something beyond his hospital rounds at Haslar. The opportunity to work again with Macgregor Laird was perfectly timed and he was eager to return to the Niger River to conduct a second expedition that would focus on trade. This time he would be designated as the mission's leader from the outset.

Because Baikie was to be in overall command of the mission, the surveying duties would fall to Lieutenant John Hawley Glover of the Surveying Branch of the Royal Navy. Glover was only twenty-eight years old but had already been in the Navy for fourteen years. He had served on surveying ships in the Mediterranean and along the west coast of Africa. He had been wounded in the war in Burma and had been mentioned twice in dispatches for his bravery. Baikie would have been delighted to have this addition to the group. His experience and fortitude would serve the mission well. In addition to Lieutenant Glover and Reverend Crowther, Baikie convinced John Dalton and Daniel May from his earlier voyage to join him for this second expedition.

Figure 10.1   The Dayspring © *The Bodley Head and Random House*

Meanwhile, the government had also entered into a contract with John Laird to provide a ship suitable for the requirements of the mission. In response, Laird had developed a new ship for this voyage called the *Dayspring* (Figure 10.1). It was similar in design to the *Pleiad*, but at 77 tons was larger, allowing it to carry a larger crew and many more goods for trading. To maximize the venture, the *Dayspring* would be followed by a second ship called the *George*. Once they had reached the junction of the two rivers the expedition would literally go in two different directions.

Where the primary exploration conducted on the earlier voyage had been along the Benue, the plan for the trip was for the *Dayspring* to go to the northwest at the confluence and continue to explore the Niger River, while the *George* would follow the path that the *Pleiad* had taken three years earlier and travel up the Benue. The captain and crew of the *George* would formalize trading agreements and establish trading and missionary sites with the contacts Baikie had earlier established. After a period of several months the two ships would rendezvous at the confluence and return together to Fernando Po.

Although the *Dayspring* would be entering new territory, it was anticipated that the string of trading sites could be continued on the upper Niger River as well. The party would travel on the river as far as navigation, supplies, and the course of the waterway allowed. Then a small delegation would travel overland to Sokoto to meet the leader of the caliphate. If the vast trading network of the Fulani could be linked to the trading stations being established on the lower Niger and Benue rivers, Laird could control all Central Africa.

With two ships on two completely different missions, this expedition would require a much larger crew than the earlier voyage. In addition to Baikie, Glover, Dalton, and May, a captain for each of the two vessels would be required. Captain Alexander Grant was selected to command the *Dayspring*, and when on the river, take charge of the trading operations. Grant seems to have been an odd choice, as completing these tasks would require a man with some diplomatic skills. Grant had at one time served in the American Navy, which was a "rough school," and afterwards worked along the West Coast of Africa among a group of "notorious characters."

At any rate, Grant was selected and would take the lead for the voyage and would pilot the *Dayspring* until she reached the confluence of the two rivers. At that point, he would take command of one of the two ships while Captain John Penman MacIntosh would become the skipper of the second vessel. Dr. George Berwick would serve as chief medical officer and would be supported by Dr. John Davis, who had been seconded from the *H. M. S. Hecla*, as assistant medical officer.[146]

The addition of Charles Barter to the crew was one that Baikie was particularly pleased to see. Barter was a botanist who had been trained at Kew Gardens in London and was currently foreman of Regent's Park, which was maintained by the Royal Botanic Society. He was widely published and was considered to be one of the top botanists of his time. William Jackson Hooker, Director of Kew Gardens, wrote in the *Kew Gardens Miscellany* (1857):

> The furthest exploration [of the] Kwora and Benue by William Balfour Baikie, R. N.—the eminent success which attended the researches of this gentleman in his survey of these waters, as described in his "narrative" lately published, having returned without the loss of a single man of his company, has induced the Admiralty to fit out another expedition for further investigations in this same region; and what was felt to be so much wanted on the last voyage, is happily supplied on the present occasion. Mr. Barter, one of the most intelligent gardeners in the Regent's Park Botanic Garden, has been appointed botanist to the expedition, and if his life and health be spared, the results cannot fail to be of great importance.[147]

Abdul Kadar, a Fulani translator and Muslim, had been appointed specifically for the planned negotiations with the Caliph at Sokoto. James Howard would serve as purser, and Robert Reese, as first mate, rounded out the crew. In all, the *Dayspring* would carry the ship's complement of eleven and what was referred to as the "government party" of five.

When the subsidy was granted and the expedition had been fitted out, Laird met the leaders of the CMS. All had agreed that a minimum of six of the trading stations would include an established church, which would be the basis of future mission work. Notices had already been posted in Sierra Leone, inviting emigrants to travel with the expedition to new homes along the Niger. More than that, Crowther and the Reverend J. C. Taylor, an

Igbo pastor, would be joined by twenty-five additional African emigrants from along the coast, who would serve as schoolmasters and evangelists in the various missions.[148] The CMS delegation would join the rest of the exploring party at Fernando Po.

This was an ambitious plan for several reasons, the least of which being that there were not yet many ordained African ministers to be found. After a great deal of effort, six African ministers were selected to accompany Reverend Crowther. The plan was that as commitments from the local chieftains were secured for the founding of the church missions, one of these CMS ministers would be left to begin the work of building and organizing each site. The selected African emigrants would be divided into six equal groups and would form the nucleus of the Christian congregation for each new mission.

Four of the CMS ministers assigned to participate met untimely deaths from sickness and disease before the party could embark from England. Another asked to be excused from the mission for reasons that were never made clear. As a result, only Crowther and Taylor would participate in the voyage. Crowther was also joined by his old friend Simon Jonas, and two youths who were in the process of studying for the ministry.[149]

MISFORTUNE, DELAYS, AND SICKNESS

The *Dayspring* left Liverpool on May 7, 1857. From the start of the second voyage, misfortune beset the expedition. The success of Baikie's earlier mission was not to be duplicated. Only three days out of Liverpool, a Prussian schooner called a barque cut in front of the *Dayspring*, causing the two ships to collide. The damage was not enough to make them turn back, but it did slow their progress and detained them for extra days at Madeira while repairs could be made. Two days later, while crossing the Bay of Biscay along the French coast, a massive, violent storm was encountered. Having already suffered major damage in the collision with the barque, the *Dayspring* was nearly sunk. The three days needed to repair the ship at Madeira were followed by additional days without enough wind to use the ship's sails. The *Dayspring* was forced to travel slowly under steam. This took an agonizing additional sixteen days to reach the start of the Gulf of Guinea near the border between present-day Liberia and the Ivory Coast.

Sickness quickly became an issue for those onboard. The health of the crew on Baikie's first voyage had been among his greatest successes. However, before ever reaching Africa many of the crew members were already suffering with bouts of fever. The extended time needed to reach their island staging area continued to create problems. Unlike the previous trip on the *Forerunner,* where the voyage out was a timely thirty-four days, it took the *Dayspring* forty-seven days to reach Fernando Po. There, they were soon joined by the schooner *George,* which had sailed from Portsmouth under the command of a transfer captain. At Fernando Po, Baikie reconnected with Reverend Crowther and was introduced to Reverend Taylor. There he was also reunited with Consul Lynslager.

While still anchored in Clarence Cove, although once again taking the prescribed doses of quinine, Baikie and most of the Europeans continued to suffer from recurring fever. This further delayed their crossing to the continent and the *Dayspring* and the *George* did not leave Fernando Po until July 1.

TRADE AND RELIGION

On July 6 the *Dayspring* was in the Bight of Benin where they crossed the sandbar and entered the tributary called Acasa Creek, which connects with the Niger River. Traveling some distance behind, the *George* found the surf too heavy to make the crossing and had to wait until the following day to enter the creek. Once the ascent of the Niger began, the two ships were lashed together to travel as one. To avoid constantly running aground the plan developed in the earlier voyage was again used. At dawn, two or three crew members would scout ahead in a small boat, marking the channel for the following day's sailing. They would return to the ship and the *Dayspring* and *George* would then set out the following morning and travel as far as the passage had been marked. In this manner the two ships slowly traversed the delta and entered the main channel of the Niger River on July 20. Once the two ships entered the main channel the *George* was towed by the *Dayspring* until they reached the confluence.

It was important that the first trading site and church mission be located close to the entrance of the river and the town of Aboh had been selected for this purpose. Crowther was fondly remembered at the places he had visited in 1841 and 1854. His reception, plus Baikie's trading success on the earlier voyage, meant the party was usually warmly welcomed by the local rulers. This was especially true at Aboh, where the old leader, or Obi, had shown such a willingness to receive the European guests. The Obi had died during the previous year and his two sons, Tshukuma and Aje, were now serving as headmen. In 1854, Baikie and Crowther had met Tshukuma, who was favorably disposed towards the mission. However, his brother Aje, whom they were meeting for the first time, presented a wealth of problems. Aje was given an invitation to meet with the group onboard the *Dayspring,* as he had expressed an interest in seeing the ship. His visit was inauspicious. Upon arriving he immediately demanded rum. During the next hour he stole, or attempted to steal Crowther's slippers, the ship's dinner bell, the cushion that had been provided for his chair, and a cigar from the pocket of the ship's captain.

With the initial ill-omened meeting concluded, and with Baikie and his men undeterred, the party landed and entered negotiations with both brothers to purchase a piece of ground for the mission and trade station. Each brother had separately received gifts upon the party's arrival and care had been taken to make sure the offerings were as comparable as possible. However, upon seeing what his brother had been given, Aje became heatedly jealous of Tshukuma's gifts. He was finally pacified with an additional pink cocked hat and matching umbrella and the site for the mission and trade station was secured.[150]

Leaving Aboh, the *Dayspring* traveled 140 miles to Onitsha, well within Igbo territory. At first, the inhabitants appeared with their weapons, but they were soon reassured and befriended and led the party along a road toward the town. Once they arrived, they were amazed to discover a community of about 13,000 and a town that was about one mile in length, with one broad road dividing it into two sections.

In the negotiations, Simon Jonas and Augustus Radillo, another liberated slave of Igbo descent, acted as interpreters. Baikie spoke for the British government, while Captain Grant dealt with commercial matters, and Crowther led the missionary group. Plans were presented and after a long and lively conference, the king stepped forth and asked the people whether they agreed to the proposal or not. There was general agreement and immediately a spot was designated where the mission buildings could be erected, and a house was rented in preparation for a factory to conduct trade.

Because this was to be a major site for both trade and the mission, and despite the lateness of the date and the deteriorating schedule, additional time was allotted for preparation. Days were spent in the village with nights spent back on the ship at anchor. One morning, Crowther was the first of the party to enter the village. There he observed great rejoicing, beating of drums, and dancing. He asked one of the headmen why they were celebrating and was told it was to honor the leader's relative, who had died some six months ago.

Simon Jonas had remained on shore the previous night and had heard that a human sacrifice was to be made in memory of the dead man. He now shared this information with Crowther. When confronted by Crowther the headman admitted that this was the plan but that the intended victim in the human sacrifice had not yet been killed. Reverend Crowther spoke earnestly and at length to the headman, and to a large crowd standing in the village square. He expounded upon the evil and sinfulness of this wicked practice. He became more fervent when it was discovered that the intended victim was a female slave.

The headman agreed that the human sacrifice would not be performed, and that a bullock would be killed instead. He proposed that Crowther should now buy the woman and that the headman would buy a bullock with the cowries Crowther provided. Crowther indignantly refused, saying he was not a slave trader. The headman then proposed that the woman should be sold to somebody else. Crowther considered this option, which he saw as a better choice than her being killed. He agreed to the plan and the villagers left in search of a buyer and a bullock. Later, when Simon Jonas returned to the ship, he reported than no one had purchased the poor woman and so, ironically, she had been set free.[151]

Because of the lack of personnel, no one had been left to supervise the proposed mission at Aboh. However, Onitsha was a much more important location and Reverend Taylor was left behind to prepare the site and to begin work on the mission. On July 31 the *Dayspring* left Onitsha and headed for Idah. After much delay upon entering the village, and while the royal court was paraded in all its finery to impress the visitors, Baikie and his men were granted an audience with the Atta, a title given to the local ruler. He received them seated on his throne and dressed in a rich silk and velvet robe of a light green hue. The negotiations went as planned this time, with Baikie taking the lead. The discussions were greatly assisted by the presence and support of the Lady Adama, the dowager queen, who seemed quite taken with Baikie. In a remarkably short period of time, a site for mission buildings and a trading station were secured in a favorable location in the village.

Despite a very difficult start, beginning with their travel from England and the ongoing concerns about the ubiquitous fever, the party had been able to establish three successful trade centers and missions. The crew seemed to be adjusting to their time on the river and all were currently healthy. Perhaps Baikie's luck had taken a turn for the better.

CHAPTER ELEVEN

JEBBA ISLAND

Leaving Idah, the ships were soon anchored at the confluence when the misfortune began again. The quinine treatment seemed to be having little impact among the crew and often half or more of them were stricken with the fever. Personnel problems were also beginning to surface. May, who had proven so valuable on the initial voyage and whom Baikie had specifically requested for this expedition, suddenly seemed unable or unwilling to work with Lieutenant Glover. The relationship evolved after May expressed dissatisfaction in his role as a subordinate to Glover to his defiantly refusing to obey orders given by anyone.

The Kru sailors, recruited on the voyage out, and who had served so bravely and adeptly on both missions, were also in a state of near mutiny. They complained bitterly about the treatment, including the physical punishment, they were receiving from Captain Grant and Dr. Berwick. Although Grant and Berwick denied any wrongdoing, Baikie demanded that they change their approach toward the Kru sailors and treat them humanely and with respect. Angered by what they viewed as an unwarranted reprimand the two men then turned their aggressive and abusive behavior toward the members of the government party. As a result, Baikie was forced to send a formal request to the Foreign Office for the suspension of May, Grant, and Berwick.[152] But beyond this official notification there was little he could do. Like it or not, these sixteen individuals would be locked together until the mission had been completed or a solution to the various disagreements and personality conflicts could be found.

The point where the Niger and Benue rivers meet marked the area as a point of great value in the work that would follow. This site had been offered by Muhammed Bello to Clapperton as early as 1824. Baikie had seen this as a prime trading site on his earlier voyage. His views had not changed. However, obtaining permission to establish a trading station would not be easy. The confluence was controlled by the Nupe people, whom Baikie was planning to visit later in the voyage. The Nupe were part of the Sokoto caliphate but their leaders were located a great distance from the confluence. There was no local leader with whom the explorers could negotiate. After much debate it was determined that discussions would have to take place when the overland party reached Bida. However, during their stay, Glover did extensive survey work in anticipation of a future trading center being located on this site.

THE DAYSPRING GOES ALONE

After three weeks spent resting in order to get the crew healthy; and after dividing the supplies, personnel, and duties; the *Dayspring* and *George* went their separate ways. Captain Grant transferred to the *George,* which would proceed up the Benue to trade. Captain MacIntosh assumed command of the *Dayspring* and those onboard continued charting and surveying as they moved up the Niger River. It was clear that Captain Grant was reluctant to give up command of either ship. He insisted that Dr. Berwick remain with the *Dayspring,* apparently to serve as his agent. This put both senior medical officers, along with Dr. Baikie, on the *Dayspring*, while the *George* entered the Benue with only the junior assistant surgeon to handle the medical needs onboard. Grant was technically responsible for all aspects of command connected with the two vessels. Baikie was incapacitated, having suffered from illnesses since arriving at Fernando Po, and now he was far too sick with fever to challenge the order. For better or for worse, division of the personnel was completed to Captain Grant's specifications. [153]

Continuing up the river, the *Dayspring* was soon at Eggan, where they found a village with an elderly local ruler who remembered the 1841 Model Farm Expedition and who received the group very cordially. Although all had been suffering, Robert Reese, the first mate, had been getting progressively weaker and was not responding to treatment. Shortly after arriving at Eggan, he died. After a short ceremony conducted by Reverend Crowther the mate was buried on a stretch of sand beach along the river.[154] The delegation moved upriver, but the progress remained difficult, and sickness and fever continued to decimate those on the *Dayspring*.

Baikie must have been unsure why the quinine preventative procedure that had proven so successful on the previous voyage had failed to prevent this sailor from contracting malaria. Although he does not discuss this death in any detail, he would have taken this loss personally and he would have taken it badly. He had surely hoped to complete this voyage, as he had the last, without a loss of life. Although his success was still far greater than any of those who went before him, he could not have seen this one death as anything but a failure. If other deaths were to follow, he would need to reconsider his theory about the use of quinine as a preventative.

MEETING THE ETSU OF NUPE

The *Dayspring* continued its ascent to the village of Wuyaji. A successful visit to the Etsu of Nupe in his new capital at Bida was critical for the mission's success and both Baikie and Crowther were anxious to make the journey there. Baikie wanted to make contact on behalf of the British government in hope of securing trade agreements. Crowther knew the support of this powerful ruler was essential for the establishment of missions north of the confluence amid the Islamic Fulani. However, before they could begin the overland journey, Baikie was again stricken with a fever. It was decided that Glover and Crowther would represent the party and Baikie would follow if his condition improved.

Crowther and Glover sent a messenger to announce their intention to visit the capital. Usman Zaki sent a return message encouraging them to come, providing horses for their travel, and indicating that they were welcome to stay as his guests for as long as they would like. Upon completion of their visit, he would again provide horses and an escort for their return to the river and the *Dayspring*. Crowther and Glover were accompanied by Abdul Kadar and the eleven-mile journey from the river to Bida was completed without difficulty.

The land controlled by the Nupe people encompassed all the territory between the Niger and Kaduna rivers. Nupe's four emirates had been involved in a series of civil wars since well before Baikie's first voyage. Usman Zaki and Masaba dan Malam had joined forces and defeated their rivals. After uniting the region, the two had agreed that Usman would be named Etsu, the chief ruler of the region. In one of Usman's first acts of leadership he had named Bida the emirate's new capital.[155]

With considerable pomp and circumstance Glover and Crowther were led into the presence of the powerful Etsu. They were greatly relieved when they were received with kindness and civility. After a full day of talk, planning, and promises they returned to the accommodation provided. The following day Baikie, who had recovered enough to travel, joined them. With the visiting delegation now complete the discussions continued. Baikie wanted to trade at the confluence but he wanted the major trading site to be established at Rabba, about sixty miles north of Bida. The negotiations remained cordial and Usman Zaki agreed to sell the land at Rabba for the establishment of both a mission and factory for trade. However, the offer ultimately was short on specifics. Their visit had been friendly and informative, but in many ways fruitless, as there was still no formal accord and the party was forced to rejoin the *Dayspring* empty handed. To make matters worse, shortly after returning to the ship one of the sailors succumbed to malaria; the second death within a week.

JUJU ROCK AND DISASTER

The expedition proceeded up the Niger River and on October 7 the *Dayspring* had almost steered safely through the narrow channel beside Jebba Island (Figure 11.1). Earlier, Glover had carefully examined the passage and had concluded that the ship's engines were powerful enough to push past the fast water. The ship was proceeding smoothly and was almost out of danger, but the forward pace was getting slower and slower. Finally, the bow turned slightly to the side and the full surge caught the *Dayspring*. She began sliding sideways with the current. Suddenly a crash shook the ship from stem to stern. They had struck a submerged rock. It was discovered later that this was an obstruction that the locals called Juju Rock and that it had stopped many native canoes trying to traverse this section of the river. The hidden boulder had torn a large hole in the *Dayspring's* hull. Disabled and leaking badly, the ship drifted for a few minutes then jammed herself upon other rocks downstream. Within a very short time the vessel's stern had settled down into the water and the ship had rolled completely on to its starboard side. Only sliding onto the rocks had prevented the ship from sinking completely into the river.

The *Dayspring* was delicately balanced on the boulders. No one could tell how long it would be before she would float clear and sink to the bottom of the river and Baikie knew

Figure 11.1 Jebba Island © The Bodley Head Random House

he had to move quickly. Local people had been watching from the shore and dugout canoes quickly came to the expedition's assistance. Baikie first supervised the successful offloading of the crew, and with the help of the local inhabitants the whole ship's company was soon safely on shore. He next organized them into work teams. Again, with the assistance of the locals and their canoes, they removed as much of the cargo as possible and transported it to shore.

While Glover and Baikie directed the securing of the cargo, Crowther coordinated making large tents out of the ship's sails. These became shelters for the stranded Europeans and Kru sailors, but the tents also created storage areas for the effects they could rescue. Before this work could be completed darkness fell. To increase the agonizing experience, they were suddenly caught within a powerful tornado. After expending all their labor to create tents and shelters following the wreck, most of the crew spent the night in the open with no protection from the rain and wind, as the powerful storm had destroyed their previous day's work.

The next morning the efforts to keep the group organized and focused were further hampered when the Kru workers suddenly and unexpectedly became insubordinate and refused to work. It seemed that without a ship they felt no compulsion to continue the agreement under which they had been employed. Baikie, outraged at the mutinous behavior, approached the headman and threatened to put him into irons until they were rescued. This put an end to the Kru insurrection. But that was not all, there was also ongoing unrest among the Europeans and Baikie determined that Dr. Berwick was once again the cause of the trouble. He immediately secured two canoes from the local people and ordered Berwick and one of the engineers back to the confluence to await the relief ship. This left Baikie, Glover,

Crowther, one of the engineers, Assistant Surgeon Davis, Dalton, Purser James Howard and Captain MacIntosh remaining near the wreck of the *Dayspring*.[156]

Macgregor Laird had attempted to provide for all contingencies. There was already a plan in place for the *Dayspring* and *George* to meet at the confluence. Should one ship not arrive the other would go in search of the missing ship and crew. Laird had also sent a third ship from England that, after three months, would follow them onto the river. The stranded party was aware that this vessel, the *Sunbeam*, was traveling behind them. They assumed that within a few months at the most either the *George* or the *Sunbeam* would discover the wrecked ship and the stranded crew. They would then be picked up and taken back to the coast. Although they were short of food and shelter, Baikie anticipated that they had more than enough supplies to last for that length of time.

The camp was quickly organized and the three principals; Baikie, Glover, and Crowther were determined that the time on the river would not be wasted. Glover secured the loan of one of the local villager's canoes and began a careful survey of the river and several of its tributaries. Baikie traveled on foot throughout the region, visiting the local leaders and entering into negotiations for future trade. He also created detailed charts and mapped the entire region. Crowther accumulated information about the four emirates that comprised Nupe and about the caravan route from Kano to Rabba. He also began an extensive study of the Fulani language.

Baikie had been preparing reports for the Earl of Clarendon, Secretary of State for Foreign Affairs, throughout the voyage. These had been collected with the intention of transporting them back to Britain by sending them overland to Lagos and then by ship to Britain. On September 28, he had completed a seven-page summary, outlining the mission to that point. A week later, the *Dayspring* had been wrecked and the exploring party had been stranded. On October 29, Baikie again wrote to the Secretary of State for Foreign Affairs with a three-page summary on the loss of the ship. These two letters, along with one written by Lieutenant Glover, would be the first communications to be carried by Glover when he traveled from the site of the disaster. But it would be some months before the overland route to Lagos could be traveled and opening up communication with the outside world could be established.[157]

By December, neither the *George* or the *Sunbeam* had arrived, and the crew celebrated Christmas Day on the banks of the Niger. From their meager supplies, one of the party was even able to concoct a plum pudding. But shortly after the festivities had ended Mr. Howard, the purser, became ill and died. This was followed within a few weeks by the death of one of the Kru sailors. Baikie had now lost all faith in the use of quinine as a preventative. The supply he had been able to salvage from the *Dayspring* was already running low and at this point, obviously disheartened, he discontinued its daily use.[158]

Perhaps he gave up too soon. Kirk and Livingstone reported a similar result while replicating Baikie's practice in East Africa. After months with everyone remaining healthy, two members of their party suddenly began suffering from recurrent chronic malaria and were quickly close to death. Like Baikie, Kirk and Livingstone were distributing the quinine mixed with wine to make it more palatable. The men may have disregarded the directions, taken a lesser dose, or simply consumed the wine. It also could have been a difference in the potency within a batch of the crystals. Whatever the reason, Dr. Kirk decided to take a different path from Baikie. Kirk confined both men to his care and increased the dosage until they began experiencing ringing in their ears; a sign of a quinine overdose. He then slightly reduced the dosage and continued administering the daily quantity. In two weeks, the two men had recovered enough to be sent back to the coast.[159] Had Baikie followed the same procedure as Kirk, he might also have completed his second voyage without a loss of life.

In January Baikie received word that Usman Zaki had died and that Masaba had assumed the title of etsu. Fortunately, the change in leadership had a positive impact on the stranded party. Following Usman's death they were visited by a delegation that expressed Masaba's desire to support the shipwrecked crew. The group brought with them a large supply of provisions and enough cowries to allow the stranded explorers to buy food locally.

By January it was apparent that something must have happened to the *Sunbeam*. The *George*, which should have returned to the confluence months earlier, had also failed to come searching for them, as had been the plan. The shipwrecked crew had to face the grim reality that perhaps no one was aware of their plight. There was an urgent need to develop a strategy or the entire group would die on the river. The first part of the plan resulted from luck rather than good management.

COMMUNICATION WITH THE OUTSIDE WORLD

Although in many ways a disaster, in one respect the sinking of the *Dayspring* had a positive impact on travel in West Africa. Groups of indigenous people were constantly visiting the camp, some out of curiosity and some attempting to provide genuine assistance. One day, amid a crowd of warriors, a strange voice saluted them with, "Good morning, sir!" The speaker proved to be Henry George, a colleague of Crowther's from Abeokuta. George had joined Masaba's army as a participant in the civil wars and was currently traveling overland to Bida. This meant that it was possible to send a messenger with letters to Abeokuta and from there on to Lagos. This would alert the outside world of the plight of the *Dayspring* and her crew.

This providential meeting led to Crowther engaging George as a guide. Letters that had been written earlier were quickly made ready and a small party, led by George, set out on their overland journey to Abeokuta.[160] This would be the first of several journeys between Jebba Island and Lagos to deliver and receive correspondence and to secure supplies. This route between Lagos and the Upper Niger would be in continuous use long after the expedition members had left the river. The great merit of the route was the fact that it was entered at Lagos and the next stage was at Abeokuta. Both places, unlike the delta and the lower Niger, were friendly towards the British.

Once it had been established that mail could be transferred overland to Lagos and then on to Britain, Baikie resumed regular correspondence between the encampment and those at home whom he thought needed to be kept aware of the expedition's progress. While most of his correspondence was with the sponsoring organizations, one piece that has survived from that period is especially noteworthy.

Baikie, having met Charles Darwin through his association with John Richardson, had been in regular communication with the noted scientist for several years. Just prior to the expedition leaving Britain, Darwin had asked for Baikie's help in corroborating or refuting William Freeman Daniell's assertion that mammals found on Annobón Island, São Tomé, Principe and Fernando Po were identical to those found on the African mainland and therefore must have traveled to these islands by crossing the deep water.[161]

Although it can be extrapolated from the language within this letter that the two had been in communication for an extended period, only one letter from Baikie to Darwin has survived and even that letter is incomplete. Darwin's initial request for Baikie's help has not been found. However, it is apparent that Darwin had been aware of Baikie's preparation for his second voyage. It is also obvious that they had spoken enough about the planning of the expedition for Darwin to knew the general route that Baikie would be following to and from his entry into the Niger River.

The surviving letter also indicates that Baikie was contacting Darwin for the first time since sailing from Britain. "Allow me to address a short note to you with reference to your communication to me before I left England last spring." It is curious that in this letter there is no mention of his being stranded following the loss of the *Dayspring*. Someone not knowing of the wreck would assume that this "encampment" was a planned part of the expedition. Baikie goes on to describe his plans to visit the islands requested by Darwin on his return voyage. He concludes by stating, that because of the depth of the waters and the swiftness of the currents, he supports Darwin's position. "[By studying] a chart of the currents...you will see how unlikely Dr. Daniell's supposition is of animals having been floated to these little islands."[162]

Baikie also took this opportunity to resume correspondence with his mentor, Sir John Richardson in a letter dated August 1, 1858. As with his correspondence with Darwin the letter begins not with a discussion of the loss of the *Dayspring* and the party's dire circumstances, but with a lengthy description of the specimens he has been able to collect. Only later, in the second page of his letter, does he mention his anticipated rescue. "Our river, since the fourth of this month, has been steadily rising, and we hope soon now to see our ship." He continues with a discussion of the favorable location of the encampment and the general disposition of the expedition members. "We are all kept well and in excellent spirits, having abundant occupation to keep us employed." The letter concludes with additional information on his collecting and his regards to Lady Richardson.[163]

Buoyed by the knowledge that overland travel was possible, Lieutenant Glover implemented the second phase of the rescue effort. Having spent the previous three months familiarizing himself with the surrounding area, and with his surveyor's knowledge of the coastal region, he decided to set out for Lagos on foot. Leaving in early February, he managed to travel overland through Ilorin and Ibadan to Lagos and then on to Sierra Leone, which was the closest point where stores and provisions could be obtained.

It was on this journey that Glover encountered a group of Hausa who had been rescued by the West Africa Squadron. When they were freed they had been released in Sierra Leone.

The men wanted to return to their homes in Kano but were afraid they would be captured and sold back into slavery. Glover agreed that they would travel together, and they all set out for the journey to Lagos. When the party arrived at Lagos, they encountered another group of Hausa who were still slaves. After talking with the Hausa from Sierra Leone, the slaves made the decision to escape from their masters and to travel under Glover's protection. The entire group was chased out of Lagos by the enraged slave owners but were successful in their escape. They had another harrowing experience while traveling through the city of Abeokuta where they eluded an armed group determined to enslave the Hausa and kill Glover. After their second escape Glover decided to train the entire group of Hausa as a fighting force and they served as able and willing bodyguards for the remainder of the journey.[164]

These initial recruits would be utilized by Glover when he became Governor of Lagos to form the beginnings of the Nigerian Army, and in 1900 would evolve into the West African Frontier Force under the direction of Fredrick Lugard. This unit received royal patronage and became the Royal West African Frontier Force, distinguishing itself in both world wars.

RESCUE AND RETURN

It was not until May that Glover returned to a very sick and despondent group at Jebba Island. The stranded party had received word during his absence that the *Sunbeam* was, in fact, on the river. However, the *Sunbeam* was a very different ship from the *Dayspring*. It required a much deeper channel and it had started too late in the year, missing the rainy season and the high water it provides. As a result, it had been unable to move far beyond the delta and would not be able to effect the rescue until the river had risen. The rainy season would begin within a few weeks and the *Sunbeam* was planning the rescue for late June or early July. Baikie made the decision that the party was too weak to travel to meet the *Sunbeam* and that the supplies Glover had provided would allow them to wait at Jebba for another six weeks. And so they waited.

The *Sunbeam's* crew had first been delayed due to their late arrival and the dropping of the river at the end of the rainy season. Forced to remain anchored on the Niger for months, they too were overcome with bouts of fever and at times they did not have enough healthy crew members to manage the ship. By the time the *Sunbeam* finally arrived it was October and the party was again nearly out of food. Only the ongoing provisions delivered by Masaba had allowed them to survive.

The rescued party traveled back toward the confluence and then on toward the delta, picking up additional members of the party that had been sent down river earlier. They also met the *Rainbow*, a second rescue ship sent out from England on its way upriver in response to the letters Reverend Crowther had sent earlier to Abeokuta. Together the two vessels returned with the remainder of the survivors to Fernando Po, arriving on November 8, 1858.[165]

Upon reaching Fernando Po, the mystery surrounding the failure of the *George* to search for the missing *Dayspring* was solved. Captain Grant had been trading on the Benue for only six weeks when the trouble had begun. Involved in constant quarrels and hostile interactions with the local people, he had finally resorted to violence. He had shot one local villager and had forced another into the river, where the man drowned. Grant was captured and taken prisoner by the villagers who tied him to a tree. They were planning to kill him, but he begged for his life and was allowed to leave provided he headed back down river immediately. The word of what he had done preceded him. The ship was under attack for the entire route to the delta and one of the crew members was shot before the *George* left the river. Upon reaching the Niger River, Grant did not even slow down but headed directly to Fernando Po, where he abandoned the *George* and boarded the first packet bound for England.[166] One assumes he never give a second thought to the fate of those he had left behind. But his actions did prove that Baikie's earlier assessment of him had been correct.

Baikie knew that Laird had been given a five-year contract with the government and he was still determined to make the commercial enterprise work. Many of those involved in the Jebba Island incident had initially accepted Baikie's invitation to try to establish a British consular agency at Rabba. However, the recent disaster of being stranded for a year had dampened their enthusiasm and now most of the participants seemed intent on returning to Britain. The whole of the *Dayspring's* crew was sent home on the *Sunbeam*, and four more of them died within a few days of sailing from Fernando Po. Grant and May returned home in disgrace and were officially suspended from the personnel of the expedition.[167]

However, five of the group—Crowther, Baikie, Glover, Barter, and Dalton—did not return to England with the others. Crowther had elected to leave the vessel at Onitsha to spend some time with Reverend Taylor and the workers there. After resting first at Fernando Po, then at Lagos, the remaining four traveled back to Rabba via the overland route that Glover had developed. They spent six months continuing to negotiate with Masaba for a trading site somewhere in the Bida emirate and planning for the long-delayed journey to Kano and Sokoto. In July, Barter died, and the three remaining members of the expedition traveled back to Lagos, where Glover soon sailed for England.

After his visit to Onitsha, Reverend Crowther had decided that he too wanted to take the overland journey from Bida to Abeokuta. He secured transportation upriver from Onitsha to Rabba, but the journey completely undermined his health and he almost died from a severe attack of dysentery. After an extended rest he set out over land to complete the 300-mile journey. He pushed steadily through Nupe country, then through his Yoruba homeland, until he reached Abeokuta. Upon reaching Abeokuta he was surprised to be met by Baikie who had traveled with Dalton up the Ogun River from Lagos. Baikie had decided that he would establish Laird's trading station even if it meant assuming the challenge alone. Baikie would be staying in Africa.

CHAPTER TWELVE

LOKOJA

It was early August 1859 and Baikie stepped carefully onto the mud-covered bank near the remnants of the Model Farm settlement. Where once there had been great hopes, now there was little more than the occasional remains of a wall or an abandoned piece of equipment. As Baikie turned back to the ship he could see Dalton and two of the Kru sailors with whom he had spent the previous year. When the *Sunbeam* had finally rescued them, everyone else had elected to return to England or to their homes along the African coast. However, these three had chosen to stay and they were already descending the gangplank with the first of their personal effects and the few goods that Baikie had secured in Lagos and which the group would initially use for trade.

Their belongings were soon stored high up on the bank and a covering of palm fronds had been arranged to protect them from the sun and rain. As Dalton and the two Kru workers began the work of sorting and itemizing, Baikie began preparation for his return journey to Bida. The captain of the ship that had brought them to the confluence of the Niger and Benue rivers had agreed to take him back upriver as far as Eggan if he could leave immediately. Baikie was soon back onboard and heading up the Niger River. After leaving the river at Eggan he would travel overland to visit the Bida emirate once again.

MASABA DAN MALAM

The junction of the Benue and Niger rivers, and hundreds of miles north of the confluence, were all part of the Sokoto caliphate (Figure 12.1). Baikie would be meeting Masaba dan Malam, whom he considered a friend, and he was looking forward to the companionship the visit would afford. He also wanted to thank Masaba for his help. Had it not been for his support the stranded crew from the *Dayspring* would not have survived the year on the river.

Usman Zaki had been Etsu of Bida when Baikie had first visited, and the purpose of that visit had been to seek permission to establish a trading center in the Nupe Emirate. Baikie had believed the best location would be Rabba, the former capital of the emirate. Rabba was already a large town and was strategically located on the trade route north to Sokoto.

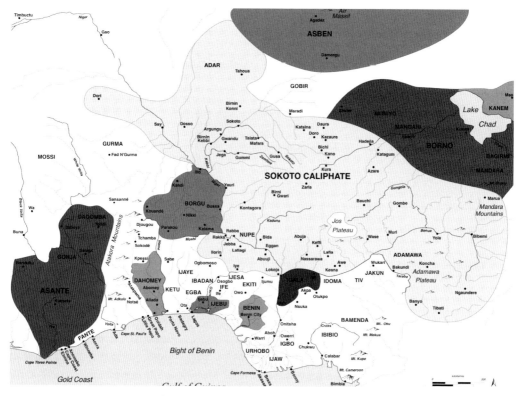

Figure 12.1 Sokoto caliphate 1850 © African Diaspora Maps, Ltd.

Usman Zaki had first agreed with Baikie's plans for Rabba. He then refused permission and finally had declined to even discuss the matter further. Shortly after this initial meeting the *Dayspring* had been torn apart on the rocks. Usman had also refused any kind of assistance for the marooned men.

However, Usman had died a few months after the sinking of the *Dayspring* and Masaba had become the new leader. One of Masaba's first acts upon assuming power had been to offer aid to the *Dayspring's* crew in the form of food and shelter. He had also given Baikie a large loan of cowries, which had allowed the distressed party to purchase additional supplies as needed, and invited Baikie to meet him to discuss further trade plans. The talks between the two of them continued for the remainder of the year and a true friendship had emerged. In addition to supporting the ship's crew, Masaba had also proven to be open to discussing the establishment of the trading station.

Although receptive to the English launching a trading site along the river, Masaba, like Usman, had been opposed to Rabba as the setting. Six months of hard negotiations had not been enough to change his decision. The two most probable reasons for his resistance were slavery and religion. Baikie's opposition to slavery was clearly known and Arab traders from the east were still using the caliphate as a base for their trade. Also, Masaba had met Reverend Crowther and was aware that the selected location would not just be for trade; it would also include a Christian mission. Rabba was positioned in the very heart of the Muslim emirate and Masaba was firmly convinced that establishing a Christian church at that site would doom the trade mission to failure.

Baikie had been encouraged instead to choose the site originally selected by Bello and offered to Clapperton in 1824 at the confluence. The area was called Lokoja. There were several reasons why Masaba believed this location would be better. Although it was still within the Nupe Emirate, on lands under his control, it was at the southern border of their territory. This meant the area included a large population of both Muslims and non-Muslims. Further, the area had already been suggested by the caliphate leadership and was currently a well-used meeting place for the diverse people of the area. Finally, since the failure of the Model Farm in 1841 others had visited the region but no group or individual had done anything to establish a permanent settlement. The few traders and the missionary who had been left in charge when the Model Farm had collapsed had continued to use the site when the Europeans left. However, in less than a year the caliphate leaders had forced this small group to leave the encampment. This left the area open for Baikie's development plan.

Baikie had always considered this location a prime trading site. However, he thought that, with the connection to Kano and Sokoto, somewhere within the Bida emirate would have proven far better. Nevertheless, this was Masaba's offer and Baikie would accept it. Masaba was the Etsu of all the territory surrounding Bida, which included the proposed site at the junction of the two rivers. Not only had he suggested the location, he had also publicly proclaimed his support for establishing a trading center at the confluence. This support would be invaluable to the project and it was not worth trying to negotiate a different location.[168]

Within Africa there were numerous kingdoms, but the Sokoto caliphate was by far the largest independent country on the continent. The caliphate was essentially a confederacy of emirates which each enjoyed a great deal of autonomy. Each of the emirates was controlled by a leader called the etsu. Each etsu reported to the leader of one of the twin capitals of Sokoto or Gwandu. A tribute was paid annually from each of the emirates to one of the two capital cities. The defense arrangements were such that the emirates were expected to defend the caliphate by defending their own immediate borders against enemies.[169]

The Bida emirate was a powerful entity. Yet Masaba reported to the leader of Gwandu and any ruling made by Masaba was subject to that leader's review. So it was necessary for Baikie to return to Bida to ascertain if the proposal had been approved and the agreement formalized. Given the often-changing politics within the caliphate, only after this meeting could Baikie know whether his trading mission was still in favor and whether he were free to begin its establishment.

The voyage along the Niger was uneventful. Baikie had enjoyed the rest that had come from being free of responsibility for the first time in years. Even the elements had cooperated, and the days had passed with warm weather and constant sunshine. At Eggan, representatives of Masaba had been waiting with horses and provisions and the overland journey to Bida had also passed without incident.

Baikie had been warmly greeted by the Etsu, who reluctantly accepted the repayment of the cowries he had advanced when the crew of the *Dayspring* had been stranded on the river. To

Baikie the repayment was more than a point of honor. It was also verification that the British and Baikie were men of principle. If trade were to be established, with Baikie as the recognized agent for England, it would be established on a basis of personal trust, honor, and friendship. It was clear to Baikie that he and Masaba enjoyed a strong bond. He wanted and needed that bond to continue and to grow along with the trading center and its associated activities.

Baikie had been given a large, dry guesthouse and two servants had been assigned to care for his needs. A series of talks had been scheduled at the Etsu's palace to iron out the details for the establishment of the trading center. Baikie and Masaba met for three days and the discussions moved from the general to the specific and were now reaching their conclusion. Baikie would be leaving the following morning. He would use the horses for his return to the river, where Masaba would provide a canoe for his to return to Lokoja. Only a few details remained to be addressed.

By the end of the day Masaba and Baikie had finalized their verbal agreement. Baikie would establish the trading center at Lokoja where "the British could stay forever." Masaba indicated he had proposed the plan to the Caliph of Gwandu and this supreme leader had also approved the endeavor. Baikie thanked him profusely and then asked: "Was there anything preventing followers of Islam from trading at the new site?" He was assured by Masaba that there was not. Although the trading station could survive on river traffic, Baikie said it would be more successful if the caliphate could also support overland trade with Lokoja. This would open trade for caravans traveling through the caliphate, in addition to the river traffic. The Etsu readily agreed and promised to begin to promote these efforts immediately. The meetings had developed beyond Baikie's expectations, and the successful future of the settlement seemed sealed.

ESTABLISHING A TRADE CENTER

By late October, Baikie had returned to Lokoja. Work on establishing the settlement continued through January 1860, with Baikie, Dalton, and the two Africans clearing the first 100 acres of land by themselves. With the salvaged timber they built houses and storage facilities, and formed enclosures. They were soon joined by representatives of the CMS and a school and church were established. As word of the settlement spread throughout the area, people came to trade and many elected to settle in Lokoja. It was soon looking more like a small city than a simple trading station.[170]

Baikie's intentions for the settlement are best described in his own words:

> The position I have now selected may possibly prove a permanent
> British Commercial site; and it is most favorably placed both as a
> convenient rendezvous and as a center for trade. A direct route to
> Nupe is now being opened, and other roads lead to Zaria and the
> interior of Hausa. As soon as I get a little settled, I shall see about
> the possibility of a direct route to Lagos.[171]

Macgregor Laird was delighted that Baikie had elected to continue the commercial operations attached to the voyage and promised as much financial and logistical support as needed. Although Baikie would not find this out until much later, changes occurring in London would also have a positive impact on his work as a naturalist and linguist.

Baikie's voyages, although financed by Macgregor Laird, had included the sponsorship of the Geographical Society of London. King William IV had consented to be a patron of the Geographical Society of London at its founding in 1830, although no charter of incorporation was granted at that time. At the start of 1859, as Sir Roderick Murchison was just ending his third term as chairman of the Society, a Royal Charter of Incorporation had been granted and the Geographical Society of London had become the Royal Geographical Society.[172] This major change in status positively impacted on the Society's work, and correspondence from the newly named organization arrived at Lokoja in the summer of 1860, pledging their continued support for Baikie's efforts and indicating their interest in the continued publication under their new banner of narratives he might elect to submit.

The selected location of Lokoja had proven an excellent choice for both trade and Baikie's work as a naturalist. Situated at the junction of the area's two great waterways, every British expedition soon recognized the trading station as a convenient meeting place for buyers and sellers from throughout Central Africa. Ships from England had a convenient single stop where they could offload their trade goods and they always found an abundant and organized supply of raw materials already prepared to be purchased and loaded. There was no longer any need to locate local sellers and enter long negotiations, as they had done in the past. This meant that the English vessels' time on the river could be kept to a minimum, and the site soon became the center of British activity for the entire region.

This location also proved to be an advantage for the Africans. With the assistance of John Glover, Baikie did establish a direct route to Lagos. He also established that the Niger was a real highway into the interior. Traders from far up the Benue and Niger rivers would travel to the confluence by canoe to trade their goods. This convenient stop relieved them from having to travel all the way to the delta to conduct their trade and it also increased their profits, by not having to work through the middlemen who were still controlling the delta.

The settlement was thriving, chiefly because Baikie had won the confidence of Masaba. The ruler had been true to his word and there were now connections between the trading post and caravan routes from throughout the caliphate. Caravans could bring goods that had been secured from as far away as Egypt. As a result, items not seen anywhere else in Africa could be found at Lokoja and more than 2,000 traders visited the settlement within its first three years.[173] This constant stream of traffic from across Central Africa and beyond also brought a variety of plants, animals, and cultural items to Baikie that he could never have collected by going out on his own.

Macgregor Laird, who since 1832 had been trying to develop trade in the interior, now began to make real money. The greatest obstacle to success—the inability of the Europeans to cope with the deadly climate—had been overcome to a considerable degree by Baikie's medical discoveries that now rendered life in the tropics less hazardous. Quinine began to displace the lancet, and the importance of hygienic practices in daily life were better understood. Baikie had also introduced a new method of trade for Europeans. Instead of having to make intermittent voyages up the river or relying on the delta traders as intermediaries, permanent trading sites could now be established at strategic points along the river.[174]

Both buyers and sellers came in large numbers from all the neighboring districts and on any given day representatives of almost every population in West-Central Africa could be seen walking the streets. The diversity of people regularly passing through the trading station allowed Baikie to study a rich variety of both regional languages and dialects. It also exposed Baikie to language variations not heard anywhere else in West Africa. Baikie's trade community gained tremendous prestige but so did his personal reputation among the local population.

THE KING OF LOKOJA

As the founder of the Lokoja settlement, William Balfour Baikie acted not merely as the informal British consul, but also as the town's doctor, ambassador, and magistrate. Masaba referred to him as the "King of Lokoja."[175] As Baikie told a friend:

> Masaba had placed the general charge of the district in my hands to keep peace, and I have already settled several disputes which threatened to lead to war and bloodshed. Natives are beginning to gather round me, as they consider themselves safe under our protection and with a little encouragement this will become a great and busy town.[176]

Herman describes this approach as

> a new kind of imperialism, a liberal imperialism, which came to characterize British rule elsewhere in the world. It involved taking over and running a society for the society's good—not by saving it through Christianity, as other European imperialisms had claimed to do, but in material terms [by creating] better schools, better roads, just laws, more prosperous settlements, more money for the ordinary people and more food on their tables.[177]

The trading station was also helping to fulfill the vision that had originally been hoped for Sierra Leone. Freed slaves, who had originally been placed in Sierra Leone or Lagos, traveled in large numbers to the trading station. Baikie settled many of them in Lokoja. Some of these educated Africans were also selected by him to help staff the additional trading posts that now stretched down river from the confluence to the delta. These centers had been created by the interest in Reverend Crowther's schools as much as in the opportunity to trade. But each of these numerous trading sites was under the control of Macgregor Laird and therefore, under the indirect guidance of Baikie.

In establishing the rules for the trading stations Baikie made no secret of his opposition to slavery. While he allowed no trading of slaves within Lokoja, nearby villages were still significantly involved in the slave trade. Taking his lead from that long-ago dinner at Cape Coast and his own experience with buying William Carlin, Baikie would periodically travel to these communities and buy as many slaves as possible. He tended to focus his purchases on the women and children, and the new settlement was soon swollen by slaves redeemed by Baikie or members of his staff from neighboring communities.

HE'S GONE NATIVE

At the time, it was common practice for slaves to be given as gifts to important guests or visiting dignitaries. When this offer was made to Baikie he always accepted. This was done to maintain ties and to avoid offending the giver; but more importantly, it provided another opportunity to offer individuals their freedom. The slaves purchased by or given to Baikie tended to remain in Lokoja, adding to its rapidly increasing population. He would first set most of them free then find paid work for them to do within the trading station or the city.

However, when Baikie could purchase or was given a child he would handle those situations differently. In most cases these children had long been separated from their families and there was no way for them to be reunited. In those circumstances Baikie often "adopted" the child and raised them as his own. Baikie began the practice of giving all the children he took in, and those who were raised by others, his surname.

It was sometime during this period that Baikie married an African woman. Whether this was a formal, legal marriage is not documented, but it is unlikely given the laws of the day. Nothing is known of her background. Perhaps she was one of the slaves that he had purchased and then freed. Possibly she was a daughter of one of the numerous local leaders with whom Baikie came into daily contact. A man of Baikie's stature would certainly have been a fine candidate for marriage to anyone's daughter. It is also reported that they had children, yet the number and sex of the children from this union is also unknown. To this day there are numbers of Africans who carry the Baikie name. There is no way to know whether they are descendants from this union, or from the children he freed and raised as his own, or perhaps those who took his name simply to honor the man.

Parsons (1999) states it was not unusual for visiting sailors and merchants to marry local women. The formal ban on sexual contact in India and other Asian colonies was not applied to Africa. It was not unusual for a British administrator or consul to have an African wife or mistress when he was assigned to a remote post. Although there was no official ban on these encounters the British government formerly discouraged the practice. But its defenders argued that "native concubines" provided sexual release for those who could not afford to marry or who were stationed in areas where there were no options other the local population. However, those who elected to live openly with a local woman were still seen as something other than British by the outside world.[178]

During this time, Baikie also began wearing traditional West African robes. Certainly, not many of his uniforms or other examples of western clothing that he had brought with him would have survived for the length of time he had been living in Africa. Moreover, the amount of his personal belongings that had been salvaged from the wreck of the *Dayspring* is unclear. Where other Europeans shunned African attire, perhaps Baikie found the traditional clothing more comfortable and appropriate to the climate than western styles of the day. Reports of Baikie's choice of dress, his African wife, his adoption of the local language, and his perceived predilection for the company of Africans over that of his countrymen were carried back to Britain by the captains of the trading vessels that traveled to the confluence, and they were not presented in a positive light.

Lady Glover, in her *Life of Sir John Hawley Glover (1897)* describes one such encounter as reported to her husband:

> On the way they stayed at Lokoja, a lovely spot at the confluence of the Benue and Kwora, a tributary of the Niger [sic], the object being to see the Consul, Dr. Baikie, and give him orders to present himself before the Colonial Office. On their arrival, Dr. Baikie came off dressed in native costume, and on returning his visit they found him luxuriously living as ruler of his kingdom under the protection of King Masaba.

Regardless of his reasons for his lifestyle, Baikie's African wife and children, and abandoning his "normal" attire were added to reports filed during his final years at Lokoja that he had "gone native."[179]

In the nineteenth century few Europeans would have referred to a "native" as a friend. They would gladly "purchase" a mistress, but they would have been ashamed to refer to her as their "wife" and she would never have been introduced into European society. For Baikie to wear native clothes, take a local wife, and study the culture, art and languages of the indigenous people was simply not done. As early as 1811, Maria Graham observed it was (at that time) best to be "outrageously John Bull."[180]

John Bull is the national personification of the United Kingdom in general and England in particular. He is a stout, jolly, middle-aged cartoon figure who first appeared in a publication in 1712. "Going native" is the opposite of John Bull and is a derogatory term for a British subject who surrenders their culture. Merchants and officials in other parts of the Empire usually developed elaborate customs and traditions to distinguish themselves from the local population. Their image was that each, regardless of their background, was a gentleman and thus superior to, and separate from, the local majority. This separation also provided a perceived legitimacy, moral certainty, and authority.

Baikie was living far away from any other European and was very comfortable with his equal status among those whom he had chosen to live. Yet those visiting him from time to time still judged him by the same standards as an Englishman who might be living in

India among hundreds of his fellow countrymen. Given the importance of segregation to the imperial system, colonial officials were extremely distrustful of any practice or institution that eroded the social distance between themselves and their non-Western "subjects." Social interaction such as dressing like the "natives," and particularly sexual relations, blurred the necessary distinction between the groups; and in Baikie, this was something that would never be forgiven by those to whom he reported.[181]

During this period, the missionary work was also being propelled forward on the success of Baikie's established achievements in trade. It was recognized by the CMS that there was no center of missionary work or of British influence comparable to Lokoja at this time. It was hoped that Lokoja would become a launching point for planting Christianity among the Muslims and pagan peoples farther north. But the vehicle of commerce that had driven the missions soon began to be the biggest barrier to their continued success.

CHARTING HIS OWN PATH

Although he was succeeding, Baikie still faced two major obstacles. The first was the lack of appreciation of his efforts by those in England. This was exacerbated by his isolation and illness. In 1859, Baikie sent a letter to Roderick Murchison, explaining why he had not been able to contribute anything recently to the Royal Geographical Society's journal:

> I work now daily, without intermission except when my time is prevented by actual sickness, from 14 to 16 hours a day, and the effects in this climate can hardly be described. I am at work by early daylight, and seldom give over till near midnight, when I am entirely exhausted. I have not been in anything like a bed for more than two years. I have had fever on more than 120 occasions, not reckoning mere recurrences after intermission, but so many distinct attacks. Secondly, I have collected information of all sorts, have offered plans and made suggestions, but they are not noticed. I receive no support, no approbation, not even disapprobation. This, sir, I think you will admit is not very encouraging. I have incurred the hostility of the slave traders on the coast, and I have been told that I am in bad odour among what may be called the Exeter Hall people, because I would not lend myself to some ultra-missionary ideas. I have had much opposition among my own people because my view of duty was looked upon as too strict. I cannot venture for a moment to compete with a Livingstone or Barth, and my duties are much more political than theirs and therefore less interesting. But if I am spared to reach home, I shall be able to show that my time has not been idly spent.[182]

Baikie's decision to establish a settlement could have not been made at a more inopportune time as far as the Colonial Office was concerned. During this period territorial expansion was not popular with the government. The costs associated with the recent Crimean War and the expenditures associated with ruling India created the perception that Great Britain was too poor to embark on additional colonial undertakings. These financial constraints were what had initially forced Britain to entrust the development of these new areas to commercial companies like those run by Macgregor Laird. Through the whole period of Baikie's time at Lokoja the spirit of conservatism and hesitation was dominant.

Baikie felt a lack of advocacy from the British government in general and the Colonial Office in particular. He was being asked to represent Britain and to operate the trading station without any financial support from the government. He was asked to perform duties and assume tasks usually conducted in a consulate or embassy. Yet, after nearly seven continuous years in Africa, he still had been given no official title or position.

He was undoubtedly also finding himself more and more at odds with the attitudes and beliefs coming out of London and directed toward Africa. Painter (2010) proposes that any nation that had participated in the slave trade found justification in a strict class system.[183] Very few in England, certainly in the Colonial Office, believed that people of African descent could be social equals to whites. Baikie, on the other hand, had long been a proponent of equality. He lived with an African woman who bore his children. He considered many of the local rulers to be among his closest friends and each considered the other an equal. He was convinced that the surest route to commercial success for both Britain and Africa was for each to see the other as an equal partner. This belief was constantly reinforced in the dispatches he sent from Lokoja to the Colonial Office and to others at home.

This is reinforced in the *African Repository* from 1892 in an article referencing correspondence written by Baikie during his last years at Lokoja:

> Dr. Baikie strongly advises an English trading station on the banks of the river Niger or Kwora, but not a territorial occupation. He describes the mind of the Central African races, even of the savages near the coast, as of an eminently practical nature, capable of appreciating the advantages of trade, and ready to turn all facilities to account.[184]

Baikie was a product of his Scottish upbringing. He possessed a quality of character that made a clear distinction between good and evil, right and wrong. Compromise was immoral and yielding a point, especially on an issue like slavery, was base. Yet he was very proud that in the entire time he had been at Lokoja he had used the threat of force on only one occasion. That had been to demand the removal of slavers who had attempted to set up shop in the trading settlement. But his feelings and recommended approach in dealing with Africa seemed to be constantly at odds with the directives and return messages coming from London. The threat or use of force seemed often to be the opening point of negotiations for his countrymen.

He was also offended by his country's attitude of European superiority. The constant message from his government clearly conveyed its belief that Africans were somehow substandard when compared with the white race. Baikie saw the two groups as equal in every sense. While the Colonial Office was only interested in the lists of items bought and sold, or in copies of formal agreements that he had initiated, Baikie wanted those at home to understand more clearly the intelligence, integrity, and honorable nature of the people he had encountered. This is reflected in the following passage from his correspondence to the Foreign Office in 1862:

> I have endeavored to record anything interesting or useful obtained regarding the little know region of Central Africa. Various tables, lists and documents, not of general interest, but at hand [can] be referred to by those concerned with such topics. I have [also] been careful to note any traditions regarding the early history of the different tribes, although at present they are of comparatively little value, the time will come when these apparently trifling stories will be sought after with as much avidity as the historical antiquary of our own country eagerly culls any legends or tales relating to those ages when our Teutonic or Keltic progenitors, barbarous and unlettered, were not a bit more advanced than many of the [African] races of the present day.[185]

But this was not an attitude shared by the Englishmen in power, or even by the European population in general. This is well illustrated in a passage from *Burmese Days,* where George Orwell's main character, Flory, is trying to convince his love interest, Elizabeth in the equality of all people:

> She perceived that Flory, when he spoke of the "natives" spoke nearly always in favour of them. He was forever praising [their] customs and [their] characteristics; he even went so far as to contrast them favourably with the English. It disquieted her. After all, natives were natives—interesting no doubt, but finally only a "subject" people, an inferior people with black faces. His attitude was a little too tolerant. He wanted her to love [the country] as he loved it, not to look at it with [her] dull, incurious eyes. He had forgotten that most people can be at ease in a foreign country only when they are disparaging the inhabitants.[186]

Another area of concern involved the slave traders from the interior and the middlemen found among the coastal people. Many African leaders continued to resist the abolition of the slave trade. Historians have suggested that the transition to non-human exports created a "crisis of aristocracy." Whereas slave-trading had been a business for "kings, rich men and prime merchants," agricultural exports profited small traders and producers.[187] The middlemen had adjusted to the abolishment of the slave trade by moving away from dealing with rulers selling their human cargo to dealing directly with the farmers who were the agricultural producers.

But, as British trade moved from the coast to the confluence, three desperate groups saw themselves as negatively affected. The Liverpool traders who remained at their factories near the delta saw the supplies that were formerly destined for them were now going no farther than Lokoja. The slave traders had been denied access to buying and selling at the confluence but faced the West Africa Squadron if they attempted to ply their trade at the mouth of the river. The local coastal people who were no longer trading as middleman saw both their status and financial standing severely curtailed. The one area where these three groups all agreed was that were all angry with Baikie and his settlement.

To combat the inland challenge that Baikie had created, these three independent groups joined forces to form the Commercial and Mercantile Association. Among their first decisions was an agreement to attack their common enemy, and they began to harass British trading ships going up the river. As a result, Lokoja was often cut off for months when steamers traveling from England were blocked or attacked as they attempted to reach the trade goods waiting for them at the confluence.

This newly formed alliance to the south consisted of people very different from Baikie's Muslim friends in the north. He describes how this association in the delta was organized:

> The members [are] the chief black and white traders in the place, and the chair [was] occupied by the supercargoes in monthly rotation. All disputes were brought before their court. If anyone refuses to submit to the decision of the court, or ignores its jurisdiction, he is tabooed, and no one will trade with him.

The Commercial and Mercantile Association could not ignore the challenge posed by Baikie and Laird's efforts. To do so would end their power and their ability to earn what had become a substantial living.[188]

MISSIONARY PROBLEMS

Despite a very zealous start and a warm welcome, the missionary work was also suffering at the hands of the river people. Part of the problem can be laid at the feet of the missionaries themselves. They tended to treat the people they were attempting to convert with utter disrespect, which angered the local communities. The missionaries showed a determination to undermine and destroy the customs and belief systems of their hosts. When they preached against indigenous practices, a typical response was, "It is the custom of our ancestors and we their children will be regarded as degenerated ones if we should swerve or depart from that which was being done from countless ages back."[189]

Most important, perhaps, is that the Christians were brought face-to-face with very strong traditionalist societies that were prepared to defend their religion and independence with courage and determination. Most of the elements of the traditional religions remained unchanged. Often, newly baptized Christians continued to observe their traditional practices, while others returned to their old ways after a few months or years. Crowther, in describing the mission work with the local people, says: "although sociable and tractable, they have no immediate desire replace their gods and ancestors with some strange and unknown saints, virgin mother or winged angels." It was at this time that the people nicknamed all Christians *ndi na ka anyi uka*, "those who backbite us." They referred to Sunday as "the day of gossip."[190]

Missionary posts were closely linked with the trading sites; and both trade and mission work had been coordinated from the outset. By this time there was not a trading station that did not have a mission and not a mission that was not located in the same complex as the trading site. The missions were deeply dependent upon this arrangement. They did not have their own river steamers, so they depended on the trading expeditions for transport up and down the river. It is not unusual to expect conflicts between the followers of different religions, but an additional complication lay in the fact that the local populations regarded all Europeans—traders, missionaries, and imperial agents—as one, and did not differentiate in their treatment of them. In the event of a disagreement with one party the inhabitants would not hesitate to analogize the rest.

Reverend Crowther saw very clearly that unless the missions could stand on their own feet their future would remain uncertain. Writing to the CMS in 1861 he stated, "If the Niger Mission is to be taken up by the Society it must be done independently of the trading factories. The natives will never believe that we are sincere until we go to work among them with earnestness and zeal." In 1862, Crowther would go up the Niger again and take with him thirty-three teachers, together with their wives and children. However, he concentrated on Onitsha and Lokoja. This very visible strengthening of the school at Lokoja only furthered its reputation as the greatest trading post in central West Africa. But it also created increased jealousy among its rivals and the attacks on the river steamers by both the middlemen and the coastal traders soon began against the settlements as well.

The attacks took a heavy toll on both the trading stations and missions located downstream, and by 1860 the attacks had brought both the trade and mission work to to a virtual standstill for all except Lokoja and Onitsha. During the months of January and February mission houses and churches were either razed to the ground or burned down. The houses of converts were also destroyed and the converts forced to flee. More disturbing to the Christian missionaries was the realization that the inhabitants of the lower Niger had gone from being collegial to being hostile. The CMS warned that "the lives of the Christians are in jeopardy every hour." Lokoja remained largely above the conflict due to Baikie's control and reputation, and because of Masaba's very visible support.

THE GOVERNMENT'S CONSUL

Then, in 1861, Macgregor Laird died. His executors decided to wind up his business in Africa and nearly all the smaller trading posts were permanently closed. Together with

Laird's death, the closing of most of the trading stations and the increasing hostilities from the slavers, middlemen, and delta ruffians, the British government now had a decision to make. Should they sever all connections with the enterprise, or officially sanction Baikie's efforts, which had proven to be commercially successful? The government had significant concerns over the reports that Baikie had gone native, and subsequently had additional concerns about his stability and capability to function as Her Majesty's official representative. The Colonial Office was also finding it increasingly difficult to deal with his attitude toward Africa and Africans, which was often so at odds with what they believed. Faced with these perplexing dilemmas, in 1860, the Foreign Office made the decision to recall Baikie and to end the Niger Expedition, and they forwarded a letter to Lokoja to that effect. These were "the orders to present himself before the Colonial Office" mentioned by Lady Glover earlier in this chapter.

When a supply ship arrived in 1861, it carried the notice of the recall, penned at Whitehall the previous year. Baikie was in no mood to submit. "I venture to defer my return," was his cryptic reply. It did not appear that the flourishing state of his settlement was appreciated in London, and he would not return until the government showed some indication of being better informed.[191] However, Dalton was now quite ill and Baikie asked that he be allowed to return on the ship to England. This request was granted, but the ship returned to England without Baikie.[192]

Baikie remained at Lokoja, but he was now the only white man left in the area and was still functioning without any official recognition from his government. His isolation made him question the point in staying on; but he had come to see his role as an ambassador for Britain and as a commercial attaché, encouraging trade links that were vital for both England and his adopted land. In Baikie's own words:

> Already traders come to us from Kabbi, Kano, and other parts of Hausa; and we are seeing regular caravans with ivory and other produce. The step I am taking is not lightly adopted. After a prolonged absence from England, to stay another season here without any Europeans, with only a faint prospect of speedy communication, and after all my experience of hunger and difficulty last year, is by no means an inviting prospect. But what I look to, is the securing for England a commanding position in Central Africa, and the necessity of making a commencement.[193]

The Foreign Office had made its decision to recall Baikie. However, backlash from merchants who had been successfully trading at the confluence soon flooded the Colonial Office. These reports not only pointed to the wealth of commerce issuing from the trading post but also listed the huge number of local African leaders "who had been won over to his views by promises of British cooperation." It was obvious, even to the Colonial Office, that closing this commercial site would have huge financial consequences. It was also clear that only Baikie had the skills and contacts necessary to make this endeavor a success.

Therefore, despite their concerns, in 1861 the British government reversed its decision. The government formally acknowledged Baikie's efforts, designated the Lokoja station a consulate, and sanctioned his leadership with the title of consul. Britain also began sending supplies and trading goods rather than having them provided exclusively by private companies, as before.

The creation of the consulate validated Baikie's position and provided him with some recognition of his efforts. It solved some personal issues that had plagued him, yet it could not address the problem of his deteriorating health. Following the deaths among the crew of the *Dayspring,* Baikie now doubted the effectiveness of the quinine regimen he established during his initial African voyage. Although quinine was available, he had discontinued its use among his staff. He had also stopped taking it himself. After the government steamer bringing both supplies and the official proclamation of his appointment as consul had departed, Baikie drove himself even harder to pursue new trade links and make further contacts. But it was becoming apparent that he was pushing himself too hard as he begun to suffer regularly from fainting spells and tremors.

CHAPTER THIRTEEN

THE JOURNEY TO KANO

Even though he was unwell, Baikie was not content to confine his activities to Lokoja. Throughout his time directing the operations at the trading center, he had regularly taken brief journeys from the confluence, staying away for days or even weeks as he surveyed the land beyond the riverside. It was late October 1861, and he now decided that he would travel on an extended journey. Kano was the principal market town in the north and was second only to Sokoto and Gwandu in importance within the Sokoto caliphate. But it was over 300 miles away and to travel there and back would take more than three months if all went well.

By 1861, through military conquest and strategic alliances, the caliphate had expanded and the Fulani were masters of most of what is today northern Nigeria and northern Cameroon. It was an advanced society that had established a uniform system of government and had generated a great revival of learning. Although the area had been united through conquest, at the heart of the Fulani revolt was an insurgency of radical intellectuals. Muslim scholars came from throughout the Arab world to be part of this society.[194]

The area around Kano generated great quantities of raw materials and produce. As Barth stated, "the great advantage of Kano is that commerce and manufacturing go hand in hand." Cloth and leather goods were produced in large quantities. Kano and its region were famed for their cloth, notably the glossy, indigo-dyed fabric that was traded in outlets as far away as Egypt, Europe, and Brazil. Corn, kola nuts, and groundnuts, salt and natron (a type of soda ash that was used in soap, to dry and preserve fish and meat, and as an antiseptic for wounds) were in abundant supply and were regularly exported. Goods that could not be produced locally were gathered in trade.[195]

Kano sat at the crossroads of four major trade routes. The route to the west, into sub-Saharan Africa, provided Kano's markets with slaves and kola nuts. The route north to Morocco, Tunisia, and Tripoli brought European goods. Trade routes also stretched northwest to Timbuktu and east to the Nile River and then to Egypt. Trans-Saharan trade routes were the longest and best known part of this system; however, there was a much greater volume of internal commerce within the surrounding savanna. A traveler wrote in 1851, "(Kano is) where everyone who pleases, and is strong enough, comes to establish himself."[196] Baikie wanted to see it all and to determine what could be brought back to Lokoja to make his operation at the confluence even more profitable.

There was a second reason for his journey. It had been reported to Baikie that materials belonging to Barth, Overweg, Richardson, and even Vogel were held in Kano. It was rumored that these materials included dozens of pages of journals, maps, and charts. The story told was that while these explorers had been staying in Kano the documents had been in the custody of a Corporal Maguire. Maguire had been killed and all the items in his possession had been stolen. For years it had been assumed that all the artifacts left in Maguire's care had been destroyed or carried off to an unknown location. It was now reported that the items had been located and that they had never left Kano. Baikie had reported this to the Royal Geographical Society and in their return message they had indicated that retrieving these materials could provide valuable insight into the early exploration of West-Central Africa and encouraged him to make the journey. The Society also expressed an interest in publishing a narrative of his travels to and from Kano should he elect to keep a journal recording them.[197]

RETURN TO BIDA

Baikie's plan was to leave Lokoja in early December 1861. This was the dry season called the Harmattan. Leaving at this date would allow Baikie time to complete his journey before the start of the summer rainy season. However, he was once again delayed due to his recurring fever. It was late December when Baikie finally left Lokoja to travel to Bida, his first stop. He felt obligated to tell Masaba personally of his travel plans. His old friend could also provide him with the resources, travel animals, and guides that would make the remainder of his journey possible.

The journey to Bida was a long trip, over 150 miles, but was accomplished without major difficulty. However, gathering all that was necessary for the next portion of his journey and finding an appropriate time to take his leave from Masaba took much longer than he had anticipated. By the time he was prepared to resume his travel it was already April. On April 12, 1862 Baikie finally started out on the second portion of his journey from Bida to Kano. He had been provided with horses for his travel and pack oxen to carry the needed supplies for the journey. He traveled with Ibrahim, a member of Masaba's court, who carried the gifts provided by Masaba to be delivered personally by Baikie to Abdulla of Kano.

Masaba also provided an escort of six mounted soldiers from Bida, who accompanied Baikie and Ibrahim for about eight miles to the next village. From that point forward, under a written directive delivered by Ibrahim, the sequence would remain the same. The group would arrive at the next town or village and their escort would return to the location they had just left. Ibrahim would locate the local leader and present the request from Masaba for an escort. Depending on the size or importance of the village or town and the local ruler's resources, sometimes foot soldiers, sometimes mounted cavalry would be provided. Sometimes it was a single escort and once it involved over fifty mounted riders. But an escort was always forthcoming and would accompany them to the next location, where the process would be repeated.

ARRIVAL IN ZARIA

By April 30 they had reached Zaria, a major market and the largest city they would encounter between Bida and Kano. The economic and political importance of the city required a longer visit and a larger tribute to the local ruler. It was discovered that the Etsu of Zaria was away, engaged in war with a group to the south. A messenger was dispatched to tell him of Baikie's arrival, which left the party free to explore the city.

Baikie had visited Zaria once before, when the crew of the *Dayspring* had been stranded on the river. On that visit he had observed the largest slave market he had ever seen, with nearly 4,000 slaves waiting to be sold. Baikie discovered on this visit that the numbers were greatly reduced and that there were approximately 300 slaves in the central market. Baikie wondered whether the slave trade was finally starting to diminish or whether the lack of numbers was simply a product of the time or the season. As he frequently tried to do, while visiting the market, Baikie purchased a young girl for the price of 90,000 cowries to keep her from being sold into permanent slavery. He planned to free her and then find her an appropriate home, work, or school.

Baikie was forced to remain in Zaria until May 20, when he finally received word that the Etsu of Zaria wished him to travel to his war camp in the south. This would take Baikie out of his way and extend the length of his journey, but he saw no way of rejecting the offer. The invitation was also accompanied by the present of a boy. So, on May 26 Baikie, Ibrahim, the boy who had been given as a present, and the girl purchased out of slavery set out toward Kaduna.[198]

The following day, after traveling only about seven miles, Ibrahim said he needed to attend to some personal business and promptly left. Baikie and the others had no choice but to make camp and wait. Ibrahim did not return until June 2. The group prepared again for travel and left the following morning. It was June 10 before they finally located the Etsu at his war camp. Baikie and the Etsu met in private conversation where Baikie presented the gifts that had been brought from Bida.

This extra travel had caused a significant shortage of the supplies that would be needed to reach Kano. Although Baikie and his group had been received warmly by the Etsu, they were told that he did not feel that it was his responsibility to provide the group with food or supplies. Instead, he gave Baikie another female slave, ironically adding another mouth to feed to a company already short of supplies. They stayed until June 15, when the Etsu broke camp to head farther south and Baikie's party resumed its northward travels. On June 18 they re-entered Zaria. They been gone from Bida nearly a month but were still no closer to their original goal.

ON TO KANO

Two days after leaving Zaria for the second time, Baikie was struck by a severe fever. It was so relentless that he was unable to travel for two days. He recovered, and the group journeyed on, arriving in Kano on July 1. Baikie thought it best to make himself as smart as circumstances would permit. He changed into his one remaining uniform before arriving in the city, hoping the military bearing would suitably impress his hosts.

Barth described Kano as a walled city of approximately fifteen square miles containing two steep hills, Dala and Kwagon Dutse that could be used as additional protection in the event of attack. Both he and Clapperton were especially impressed with the construction of the wall that completely encircled the city, which was described by both explorers as "the most imposing piece of workmanship in this quarter of the world."[199]

Upon entering, the party discovered a municipality of some 30,000 to 40,000 inhabitants. But, aside from the large population, Baikie found the place in other respects a bitter disappointment (Figure 13.1). The square clay houses, roofed with the trunks of palm trees, were scattered in untidy groups between large stagnant ponds of water. An extensive swamp, called Jakara, cut the town almost in half. The waterway was covered by reeds and frequented by wild ducks, cranes, and a filthy kind of vulture. They approached the palace to deliver the letter of introduction from Masaba, only to find that Emir Abdulla was also conducting war away from the city. As in Zaria, a messenger was dispatched to tell him of Baikie's arrival. The party was again forced to wait for the return message.

Having crossed the swamp by the marketplace to deliver a letter of introduction the party now had to re-cross it to the other side of the town, where a house had been allocated for them. Unlike the housing that Masaba had provided in Bida, this was a gloomy hovel of two floors with tiny rooms. Little holes had been cut into the side walls, but they admitted only

Figure 13.1 Kano © Hotels NG

a glimmer of light. There was little air circulation and the interior was dark and grim. With no other options the exhausted travelers settled in to wait for the messenger to return from Abdulla's war camp.

While Baikie found Kano itself a dispiriting place he thought its market was the best regulated he had seen in Africa and its bustle of activity afforded endless hours of interest. The prices were fixed by the official in charge, and if purchasers subsequently found the goods they had bought were of inferior quality, the price they had paid would be returned to them. Bands of colorful and noisy musicians paraded up and down, attracting the attention of passersby to booths in which were displayed all manner of goods. The range of merchandise offered for sale within the neat rows of bamboo stalls was quite remarkable and was in stark contrast to the rest of the dirty and disarranged city.

There were stalls selling food, silks, and ornaments. Tables held armlets and bracelets of glass, coral, and amber. Merchants offered silver trinkets and pewter rings, next to scissors and knives of exquisite workmanship. There were stalls selling make-up with indigo for dyeing the hair and eyebrows and others selling the flowers of trees for coloring lips a deep blood red. There were also items imported from abroad. There were sword blades from Malta, coarse writing paper from France and green English umbrellas. Of course, there were slaves who were being examined with the utmost care by their intended purchasers. Each slave in turn was required to show their teeth and tongue and made to cough. Although he was sickened by the presence of slave trade, Baikie believed there was much of what he saw in the way of order and structure that he could take back with him to Lokoja and apply to his own trading post. That knowledge alone would make the trip worthwhile.[200]

Days passed with no word from Abdulla, leaving Baikie ample time to continue his exploration of Kano. When not in the market Baikie occupied his time watching the women spinning cotton, dyeing cloth, and tanning skins. He was also entertained by the snake-charmers who made their serpents, some of them over six feet long, perform a kind of dance to the beating of a drum, coiled them around their necks, pointed their fingers at them, exasperated them, raised them into position as though they were about to strike, and then made them retreat by spitting in their faces. As a grand finale they would always pick up the snakes and hurl them among the spectators, a practice that Baikie never fully came to appreciate.

While in the house that had been assigned to him, Baikie received an almost continuous flow of visitors. He was delighted that there was great interest from the people in obtaining copies of the Psalms which he had earlier translated into Arabic. They urged him to return at another time with full Bibles in Arabic. This response cheered him, but other visitors told him they had been warned of the treachery of the English by the Arab merchants in the marketplace. Baikie knew that these traders saw him as a rival and were anxious to ensure that their own commerce did not suffer. There was an additional problem presented by this constant parade of callers. Baikie was now suffering from a recurrence of fever and this endless stream of curious visitors gave him no time to rest and recover.

ABDULLA'S WAR CAMP

Finally, on July 8 a messenger was sent from Abdulla with a note to have Baikie brought to his camp. Baikie was still quite ill but he was also aware that he was being continuously judged on his actions and behavior. He was the only non-Muslim that most had seen, and the reception given to those who might follow would be based on how he was perceived. Therefore, he would have to travel at once. Baikie had also read the accounts of Oudney and Clapperton and the importance of honoring local faith and tradition when visiting within the caliphate.[201] So before leaving Kano, Baikie observed the Islamic custom of Zakat. He purchased two bullocks and paid to have them slaughtered. He then arranged to have the meat distributed among the poor of the city.

Baikie's fever was a significant handicap as they left the city, but he could delay no longer. He was once again in uniform and they were soon met by the escort of some 150 horsemen with drummers and trumpeters sent from Abdulla. The guides advanced towards him at a full gallop, bidding him welcome in the name of their master and thereafter riding so close to him and in such dense masses around him that he was almost suffocated by their dust. Baikie was aware that this extraordinary respect was paid to him as the servant of the Queen of the white men, as he was described in his letter of introduction that had been prepared by Masaba. This was also evident during the first discussion between the Abdulla and Baikie.

Wasting no time Abdulla asked, "Would the English Queen provide me with guns and rockets?"

"Yes," replied Baikie, "if in return you agree to put down the slave trade." Not surprisingly, from that point forward, discussions involved only the proposed trading partnership and the lost materials from Barth's expedition.[202]

Concerned with the passage of time, Baikie pressed for information on the materials related to Barth, Vogel, and the others, beginning with his initial meeting with Abdulla. Baikie learned that the lost materials did exist and was provided the location where they could be found in Kano. Eager now to have the materials in hand, Baikie elected to return to Kano at once, promising Abdulla that he would stop for a prolonged stay upon his return.

Going back to Kano, Baikie easily located the house where the papers were allegedly kept, although it was unclear why they were at this specific location. It was certainly not a house of unusual size or appointment and the owner did not appear to be someone of wealth or power. Baikie was warmly greeted and the materials were presented without question. Sadly, the find was a disappointment. The "lost papers" turned out to be two astronomy books, written in German, and bearing the names of Overweg and Vogel. According to the owner of the house, after Maguire's murder a musket, pistol, and the manuscripts were all taken to Zinder, where they remained. Baikie was tired, ill, and running short of supplies. Zinder was over 150 miles north of Kano. He did not have the time, energy or resources to take the additional weeks that would be required to continue the search in Zinder. And there was no verification that the missing items were still being held at this location. He decided to begin his return trip to Lokoja on the following day.

True to his word, Baikie returned to the war camp for a lengthy visit to Abdulla. He remained in his company until August 14, when he began the journey back to Bida. The trip had taken far too long, and it was now well into the rainy season. Traveling was painfully slow, and many detours had to be taken because of swollen streams and rivers. But in spite of his illness and fatigue, Baikie continued to collect seeds and plants, which he forwarded to Kew upon his return to Lokoja (Figure 13.2 *see colour section*). Finally, on November 6 the party located Masaba, who was also then at war and camped near Rabba. Baikie wanted to complete his travels as quickly as possible. Unfortunately, he was so ill by this time that it was not until December 22 that he was able to set out by canoe for his home at Lokoja. He had been gone a year, traveling nearly 700 miles on foot, on horseback, and by canoe. He was debilitated and weak, and he would never completely recover from this journey.

CHAPTER FOURTEEN

GOING HOME

It was late January 1863, and Baikie was continuing to recover from his trip to Kano, constantly battling with bouts of fever. In addition to the increased hardships of his trip, he was exhausted from the constant pressures of being solely in charge of the Lokoja trading settlement. He knew that he would never recover his health unless he was able to have complete rest. He also knew that this could happen only if he left Africa. In addition to his own health concerns, upon returning from Kano he had received word of his mother's death. He had also been notified that his father was in poor health. He needed to go home, and he wanted to go home.

However, he also wanted to make certain that all he had built at the trading post would continue in the direction he had established. To that end, he began an extensive correspondence with Lord John Russell, the Secretary of State for Foreign Affairs. Baikie had already established a relationship with the Earl of Clarendon when he was Secretary of State for Foreign Affairs. Lord Russell had been appointed to this position since Baikie had established the settlement at Lokoja. Perhaps Baikie felt the need to provide him with the same information as he had given his predecessor. Whatever his reasons, Baikie began a prolific correspondence about his efforts and his suggestions on the direction that Britain should take toward Africa.

Before he had left for Kano, and while he was on route to the city, Baikie was in almost constant contact with leaders in England. Between February 13 and April 9, 1862 Baikie wrote to the Secretary of State for Foreign Affairs twelve times. He also forwarded thirteen pages of reports and corrections of errors made by earlier explorers like Livingstone and Barth. The topics ranged from the trade value of British goods to growing cotton along the Niger, and from the collection of ivory and palm oil to Baikie's recommendations for the best strategies for trading with Africans. Although these letters were all written at different times, their delivery to Britain was contingent upon the packet ships and their schedule. Thus, all twelve letters arrived at the Foreign Office on the same date, August 11, 1862 before Baikie had returned to Lokoja.[203] There is no record of Lord Russell's response, or whether he responded at all.

REQUEST TO RETURN HOME

Shortly before Baikie had begun his first voyage and as part of the government charter, Laird's company had established a system of regular communication with the West African Coast. Three ships *Faith, Hope,* and *Charity* had been built to run as regular packets to deliver mail. In 1852 John Laird, through his established shipbuilding firm at Birkenhead, had launched the 894-ton *Faith* and a ship of equal size christened the *Hope.* In May 1853, the 1077-ton *Charity* had been put into service. Thus, the uncertainty of relying on communication transported on returning palm oil trading ships or the unpredictable schedules of sporadic packet ships had been replaced by a regular, monthly delivery of mail.

However, the conflict in the Crimea had prompted the Crown to take advantage of the language of the charter, which said that the directors had to make the ships available to the government in time of need. These packet ships had been used to transport troops between England and Turkey and had only recently been returned to the purpose for which they had been designed.[204] Had they been available when the party had been stranded on the river, the suffering of the crew might have been greatly reduced. However, they were now back in service,and when the *Hope* left Africa for the return home she carried Baikie's official request to be temporarily relieved of his consular position and to be transported back to England.

While waiting for his replacement and transport home, Baikie spent the time organizing the vast number of writings and translations that he had completed since first settling at Lokoja. While he was at the station he had written a series of publications on trade and the local population, compiled vocabularies of over fifty local languages and dialects and translated large parts of the *Bible* and *Book of Common Prayer* into the Hausa language. He had also completed translations of portions of these books into Igbo, Fulani and Arabic. His *Observations on the Hausa and Fuifuide (i.e. Fula) Languages* had been privately printed in 1861, and his translation of the Psalms into Hausa would be published by the Bible Society in 1881.[205] "Notes from a Journey from Bida in Nupe, to Kano" would be published in the *Journal of the Royal Geographical Society of London* in 1867. Other translations of various African languages would be incorporated in Reichardt's *Grammar of the Fulde Language,* which would be published in 1876.[206]

The collecting, cataloging, and placing of his various finds in British cultural institutions also continued during this period. The major beneficiaries of his collecting efforts were Kew Gardens for the plant specimens and the British Museum for the discoveries he made among the rare and hitherto undiscovered animal species. The *Specimens of Shield Reptiles,* produced by the British Museum in 1873, is a document listing and describing discoveries of new species that had been donated to the museum. Next to each listing is the naturalist credited with the discovery. The list of new discoveries credited to Baikie extends over two pages and includes only the reptile species he had encountered.[207]

SIERRA LEONE

Although his request to be relieved went out in a timely manner, the government's response was slow. It was not until August 1864, with the arrival of *H.M.S. Investigator* under the authority of Commander Sands, that Baikie could finally leave for England.[208] By then he was suffering from almost constant fever. He was thirty-nine years old. Since the wreck of the *Dayspring* and the opening of the trading station at Lokoja, he had been living in Africa for seven continuous years. His travel to Kano had been the final assault on his body and he was nearing total exhaustion. His plan was to return to England and file the reports that the Colonial Office had been asking for. He would then take the time necessary to allow himself to recover fully and do nothing more strenuous than organizing his papers and collections. Then, when he felt well enough, he would return to Lokoja—perhaps in about a year. During his absence, Lieutenant Bourchier would serve as Baikie's temporary replacement, having traveled with the *Investigator* on the outbound journey.

It is clear from Lieutenant Bourchier's report that he, too, had heard the stories of Baikie's transformation into a native. His report, which was reproduced in 1867, states:

> We were all anxious to arrive and see Dr. Baikie, he being the only white man living in these distant regions. Some evil-minded persons had represented him as a niggardly, penurious being, who was quite in his element living in a half-and-half sort of civilization; and had reported that he had turned Mohammedan; that he kept a harem of women; in fact, we all expected to see a dried-up, wizen-looking piece of humanity. But our ideas were agreeably changed when the boat came alongside, and in a moment on the deck stood the man. Dr. Baikie was dressed in the cool loose country style, and very well he looked, both in the point of the picturesque, and in health and strength; he was powerfully built and was well tanned by the sun.[209]

It is clear that Baikie had adapted to his surroundings. It is also apparent that the reasons for his behavior made perfect sense in this setting and were very different from what had been reported to those in England who had characterized him as going native.

Baikie would not be traveling from Lokoja alone. He had selected several children to accompany him. Three of the them—Aaron, Joseph, and Harry—bore his name. Because of their ages they could have not been his sons. All these children were probably among those he had purchased out of slavery. At any rate, they would be accompanying him on the voyage as far as Sierra Leone. He had planned for the education of "his children" and he would be enrolling them in the boarding school at Freetown, where he had placed William Carlin. The *Investigator* was a small paddle survey vessel built to ply the African rivers. At only 121 feet in length and 16 feet across, she was not designed to carry large numbers of passengers. Commander Sands had protested at the addition of the children, saying that the ship did not have the capacity for that many additional passengers. However, upon Baikie's instance, Sands had been forced to relent.[210]

Perhaps it was the understanding that he was finally headed home, but as the *Investigator* began its journey from Lokoja, Baikie felt as well physically as at any time he could remember, and the journey from the confluence to the Niger delta passed without event. However, once the ship reached the open ocean he again began experiencing the headaches and bouts of fever. The ship's medical officer had been treating him, but nothing seemed to give him relief.

The *Investigator* reached Sierra Leone in late September and Baikie again sought out Charles Heddle, the Orcadian friend who had helped him with William Carlin. Baikie was both proud and pleased to know that William, now sixteen, had finished his schooling and was working with Heddle in his shipping business. Charles Heddle again readily agreed to monitor the schooling of this latest group of children. Baikie placed them in the boarding school and arrangements for their ongoing enrollment were soon completed.

Heddle invited Baikie to stay with his family while the *Investigator* was in port and Baikie gratefully accepted the offer. After a few days of rest in a comfortable home, he was feeling so much better that he announced to Commander Sands that he planned to remain with the Heddle family in Sierra Leone for an extended period. The captain of the *H.M.S. Investigator* was torn about how he should respond. On the one hand, he had been directed to bring Consul Baikie back to England. But he could not stay in port for a prolonged period, waiting for his charge to decide he was finally ready to make the journey. On the other hand, Dr. Baikie seemed to be recovering and having him truly healthy enough to travel later seemed to be the better course to follow. So the transport ship remained in port only long enough to discharge the passengers and to take on supplies. The captain then reluctantly bid Baikie "good health" and the *H.M.S. Investigator* left for Portsmouth, with the person he had been sent to retrieve remaining in port.

Baikie adhered to his original plan of resting and sorting out his manuscripts and natural history collection in preparation for his return to Britain. He continued his stay in Sierra Leone for six weeks with no serious setbacks. In fact, things were going extremely well. The bouts of fever, although they still occurred, were coming less often and seemed to be less severe when they did strike. In early December, he was enjoying helping the Heddle family prepare for the Christmas holiday. It promised to be a true celebration, with William Carlin taking time from work and his other children spending their school holiday with Baikie and the Heddle family.

Amid this productive and festive environment, a positive feeling prevailed. So, it was to everyone's surprise when on December 10, 1864 Baikie went into total collapse. Medical treatment was provided but nothing could stem the tide of the deadly fever and in two days he was dead. William Balfour Baikie, the man who had proven that quinine could prevent the contraction of malaria, had died from a tropical fever. Although his death was a surprise to others, Baikie may have had a premonition of it. In a letter to his sister, Eleanor, he says, "I would give all I possess for five minutes with you in the old home." It is believed to be the last letter he ever wrote.[211]

In the same way that several dates were listed for Baikie's birth, many more were listed for the date of his death. The *London Illustrated News* reported that he died on November 30. The *Orkney World* listed December 12, and *The Orcadian* initially listed December 10, 1864. His "Certified Extract of Death" indicates he died on December 12 at 3:45 pm. He was residing at that time at the Heddle residence on Water Street in Freetown, Sierra Leone. It indicates that Charles Heddle and Arthur Montague, who was perhaps the attending physician, were present at the time of his death. Baikie is listed as being thirty-five years old; although, he would have been thirty-nine on that date. He is listed as a surgeon in the Royal Navy and his cause of death is listed as "debility," a commonly used medical term of the time referring to muscle weakness and rapid weight loss.[212]

Earlier, Baikie had asked that in the event of his death he be buried in Africa. So, with full military honors he was laid to rest in the ancient cemetery in Freetown, Sierra Leone. It is reported that his funeral was the largest held up to that time. Word of his dying took six weeks to reach England and the official notices of his death were not published in the London papers until January 28, 1865. Word of his passing took even longer to reach family and friends in Orkney.

THE FALLEN SON

When word reached Kirkwall of his death, the entire community was devastated at the loss of their beloved native son. It was quickly decided that a monument should be erected somewhere in the town to honor Baikie's life and work. A memorial was paid for by public subscription and was placed in the nave of Kirkwall's St. Magnus Cathedral (Figure 14.1, *see colour section*). It was designed in the style of the thirteenth century, with three recessed arches in each side and one in each end. The arches contain the England, Scotland, and Orkney shields of arms; and the crests of the Baikie, Traill, and Hutton families. The main portion of the vault is of Orkney freestone in two colors, and the detached shafts are of Shetland serpentine. This impressive tribute from the people of Kirkwall is dedicated to Dr. Baikie "as a token of their respect for his character, talents and virtues, their admiration of his useful life, and sorrow for their own loss in his early death."

The full inscription reads:

> William Balfour Baikie, MDRN, FRGS, FBS, FSA (Scot). Born at Kirkwall 27th August 1825. The Explorer of the Niger and Tchadda, the translator of the Bible into the languages of Central Africa, and the pioneer of education, commerce and progress among its many nations. He devoted life, means and talents to make the heathen savage and slave, a free and Christian man. For Africa he opened new paths to light, wealth and liberty—for Europe, new fields of science, enterprise and beneficence. He won for Britain new honor and influence, and for himself the respect, affection and confidence of the local leaders and their people. He earned the love of those whom he commanded, and the thanks of those whom he served, and left to all a brave example of humanity, perseverance, and self-sacrifice to duty. But the climate, from

which his care, skill, and kindness shielded so many, was fatal to himself, and when, relieved at last though too late, he sought to restore his failing health by rest and home, he found them both only in the grave. He died at Sierra Leone on 12 December 1864 (Figure 14.2, *see colour section*).

CHAPTER FIFTEEN

A FRAGILE LEGACY

In his annual address to the Royal Geographical Society of London in 1863, Sir Roderick Murchison, now in his fourth term as president of the Society, spoke glowingly of the work Baikie had done representing the Society and alluded to the prospect of his return to Britain. A year later he delivered Baikie's eulogy to this same group. After providing a summary of Baikie's two voyages and subsequent work at Lokoja, Sir Roderick concluded the eulogy with the following remarks:

> I am aware that the Foreign Ministers of this country, past and present, have been well satisfied with the efficient services of Dr. Baikie; but I regret to say that some time must elapse before the real value of those services can be known. Whilst our deceased member made but scanty communications to us, he kept, as I understand, numerous journals and researches, from which he doubtless intended to compose a complete work had he not been unhappily cut off at Sierra Leone. [It is not his writing, but the friendships] which he made with various chiefs, the moral influence which he exerted over them, and the good-will if not friendship of the natives which he acquired (his messengers and people traveling in perfect safety from Lokoja to the sea-coast), are the best tests of the value of his kind and conciliatory but firm and judicious conduct. The Orkney Islanders may well be proud of having produced such a man, and I trust that the right feeling which has guided his friends to erect a monument to his memory in his native town of Kirkwall will induce Her Majesty's Government to honour his services by aiding his bereaved family with a befitting recompense.[213]

There is no record of the government providing any compensation to Baikie's family. However, Masaba was informed in a personal letter from Queen Victoria of his friend's death. The Queen promised Masaba that she would replace Baikie with another consul and did. But the consular post was withdrawn three years later and even the Reverend Crowther was then brought back from his mission station.

THE FORGOTTEN MAN

On his first voyage, William Balfour Baikie was propelled into a position of leadership as a result of a war half-way around the world, the death of Consul Beecroft, and the total incompetence of the *Pleiad's* captain. Whether he had been born to lead, or was forced into leadership, he assumed command and achieved success well beyond what any others to that point had attained. Following that successful first voyage he returned to a hero's welcome but was then passed over by the sponsor of his expedition for all their major awards. Had he been honored with one of the Geographical Society's top prizes this public recognition would perhaps have opened up new opportunities and taken his life in a different direction. However, it is just as likely that his heart was already set on a return to Africa and that the honor would have changed nothing.

On his second voyage, although it was fraught with disaster, a young and resourceful Baikie kept the party together for over a year under the direst of circumstances, until they could be rescued. Following that voyage he elected to remain in Africa, rejecting an offer of rescue provided by the government. He may have remained because he perceived himself to be a failure and did not want to face those at home. He may have delayed his return out of a sense of duty. Macgregor Laird had put his faith in him, and the job was not yet completed. Or he may have stayed because he had found a home among a people he had come to value and respect.

For whatever reason, he continued at Lokoja, creating a system of trade that became a model that would be used by the British from that point forward. In the process, he succeeded in exploring vast areas of what is now central Nigeria, as well as traveling the caravan routes established throughout the Sokoto caliphate. Those who followed were able to extend this exploration and trade because of his proof that the use of quinine prevented malaria. He was a prolific writer and linguist. He was also one of the most successful naturalists of his time. He published two books on the natural history of the Orkney Islands before his twenty-third birthday. The objects and plants that he collected during his expeditions were added to the collections of the Royal Botanic Garden at Kew, the Royal Botanic Garden at Edinburgh, and the British Museum. His work resulted in having a genus of plants, *Baikiaea* named after him. Yet, for all this success, William Balfour Baikie's life has generally been relegated to the back pages of history.

A PROGRESSIVE COLONIAL MODEL FOR TRADE

Baikie elected to remain in Africa for seven years and to fulfill the original charge of his mission. He engaged and befriended the local people and established a successful trading settlement on behalf of Macgregor Laird and the British government at the confluence where the Model Farm had failed. The contracts and treaties that Baikie established with the people of the region opened up the entire area to British trade. His work in establishing the trading station at the confluence of two great rivers became the city of Lokoja. The development of his successful trading model at Lokoja, and his demonstration that the Niger River could be the highway to open up all West Africa, set the stage for Britain and for all who would follow.

He lived and worked among the local people. Working alone, Baikie established strong ties within the Islamic caliphate. His adaptation to local culture, language and dress, and his union with an African wife allowed him to succeed within his environment where others had failed. Yet these very factors almost certainly caused him to be viewed negatively by his countrymen as having gone native, and perhaps even insane. This, coupled with the contrary stance he maintained between himself and the Colonial Office over their approach toward Africa and Africans, may have caused some of those at home to be relieved at his death and more than happy to let him fade from memory.

In establishing Macgregor Laird's interests along the Niger, Baikie also established a consulate and secured a political and economic victory for Britain in the process. Lokoja became a market that was extensively used by traders from throughout West Africa. Peace prevailed in its immediate vicinity and trade routes to Nupe, Sokoto, Bornu, and Lagos were opened. This pioneer of empire was respected by the locals for his strength of character and his real moral superiority as much as for his sympathetic understanding of the Africans and their point of view.[214]

Macgregor Laird said, "moral force in Africa was a 32-pounder, with a British seaman behind it." Baikie elected not to use force. Yet, he maintained the honor of England. He lived not only in safety but with the respect and esteem of the of the people with whom he interacted. Baikie's strength lay in his strong moral character which is reflected and described in the rules of conduct he prescribed for his successors:

1. Always strictly keep faith and promise.
2. Except for some definite purpose avoid interference with Native customs.
3. Learn as fully as possible all Native customs, etiquette and politics; but as quietly and unobtrusively as possible.
4. Learn the local language.
5. Do not be too intensely European, either in habit or dress.
6. Try to show a real moral superiority over the Natives; never

betray the least feeling of fear; keep up your own position, and never relax the discipline among your own people.

7. Never allow a theft or an insult to pass unnoticed but insist only on an apology and reparation.

8. Learn and study the characters of the principal chiefs of your district or region, and approach them accordingly. Make their friendship, but never excite the envy or jealousy of one against the other.

9. Always keep good friends with the powerful chiefs.

10. Study the character of all your intimates and acquaintances that you may know how to use them most suitably.

11. In trading matters, be patient.

12. Do not be too much shocked at the numerous untruths daily told, especially in trading.

13. Avoid commencing religious discussions, especially with Moslems. Give the Prophet credit for what is good in their creed and practice.

14. Never give any countenance to any form of heathen worship.

15. Never pay any undue or excessive respect to the chiefs or kings below the Confluence. Threat them with ordinary courtesy, nothing more.

16. As far as possible, keep your temper and study patience.[215]

Not surprisingly, those sent to follow Baikie failed to utilize the humanistic approach that had made him so successful. They maintained their distance, their English dress, language and customs. Local leaders whom Baikie had considered friends were now treated as inferiors and with disdain. Consequently, the people of the region became increasingly hostile toward the leadership of the trading station. Without Baikie's enlightened approach, and his ability to serve as a buffer between the local population and London, the governmental approach to the people of the Niger changed as well.

A sad example of this downturn is seen in 1861 when Mr. Henry, a British trader, seized oil belonging to an Itsekiri trader. The Itsekiri people live in the Niger delta and were among those making a living by serving as middlemen. This Itsekiri trader owed Henry six puncheons of oil but Henry, in a classic colonial show of power, seized fourteen puncheons. The trader asked for payment of the additional eight puncheons but was instead put into chains on Henry's orders. When he was released, he organized a raid on Henry's factory and stole a variety of stores. Mr. Henry appealed to the Consul, a Mr. Freeman, who found Henry entirely responsible for the problems. Although the Itsekiri trader was found not to be responsible for any wrongdoing, Freeman announced, "such an outrage could not be allowed to pass," and the Itsekiri trader was fined thirty-five puncheons of oil. Mr. Henry, who had been found guilty in a court of law, paid no fine at all. This was the type of British justice that finally led to military conflict following Baikie's departure.

Baikie arrived with an open mind and, working within the culture and norms he encountered, had been warmly welcomed by the people along the Niger River. The Europeans who followed seemed to arrive with preconceived notions that Africans were barbaric. The Africans, in turn, felt betrayed, and perhaps they were. Their entire way of life and system of beliefs were being ripped apart. The missionaries had served only to separate the population and to undermine traditional beliefs. The merchants who arrived after Baikie established rules that continued to eliminate the middlemen; however, they also positioned themselves in a way that maximized their profits at the expense of the locals at the confluence.

In 1868, the hostility of those who had worked so successfully with Baikie reached a climax and they killed the final Lokoja consul, George Fell, the last of a series of individuals who had been sent to replace Baikie. In 1869, shortly after Fell's death, the British elected to close the consulate because they were unable to find a man who could successfully fill the void left by Baikie's death. Expansion in West Africa was greatly reduced and British government action in the Niger River valley was reduced to a minimum.

Baikie accomplished his vast commercial success without the use of force. He lived at peace within the largest Islamic country on the African continent. Forty years later the British were able to control the Fulani only through all-out warfare and the use of modern weapons. Carland credits Fredrick Lugard with establishing "British control over the whole protectorate by a successful campaign against the emir of Kano and the sultan of Sokoto."[216] Baikie had accomplished the same feat forty years earlier through honest negotiations, respect for the local leaders, and sincere comradery with the people. Lugard has been extensively studied and his work documented and recognized in many publications. He was knighted and remains a major figure in the history of Africa, while Baikie's accomplishments remain an obscure footnote to this history.

PROVING THE USE OF QUININE

Christopher Lloyd states, "Throughout this period European travelers were handicapped by deficiencies in tropical medicine in an area where conditions [were] the most lethal in the world." Less was known of the trials and tribulations of the early explorers of West Africa because few of them lived beyond their initial explorations.

Baikie's proof that quinine could be used to prevent malaria broke new ground and despite his setback on the second voyage it was ultimately proven to be the most reliable prophylactic of its time. Alexander Bryson's article, suggesting this preventative procedure, was published shortly before Baikie sailed for Africa on his initial voyage. His findings, in what constituted a controlled study, validated Bryson's hypothesis and Baikie is given credit in the Royal Naval Museum archives as the first to test Bryson's theory. However, recognition for proving the hypothesis never appeared anywhere beyond the archives. Bryson went on to serve as honorary physician to the Queen and eventually became director general of the Naval Medical Service, while Baikie's demonstration of this revolutionary idea faded from memory.

In *The Greatest Benefit to Mankind: A Medical History of Humanity* Roy Porter said: [217]

> [Q]uinine also played a significant role in the colonization of
> Africa by Europeans. The discovery and use of quinine had been
> said to be the prime reason Africa ceased to be known as the
> "white man's grave." It was quinine's efficacy that gave colonists
> fresh opportunities to swarm into the Gold Coast, Nigeria and
> other parts of West Africa.

It remained the antimalarial drug of choice until the 1940s, when other drugs such as chloroquine replaced it. Porter states, "The [cinchona] bark was first dried, ground to a fine powder, and then mixed into a liquid [commonly wine] which was then drunk. Large-scale use of quinine as a prophylaxis started around 1850." The method he describes reflects Baikie's administration of the quinine and the date corresponds to his voyage of 1854. It will never be known if this is a reference to Baikie, as Porter does not credit this work to any individual.

The findings from Baikie's experiment were not accepted by the medical establishment until years after his death. The inability of the medical establishment to adopt new treatments, even when the improved methodology proved vastly superior to the customary treatment, was not limited to quinine and the handling of malaria. Earlier, 1,100 members of the Seaforth Highland Regiment had undertaken a five-month trip from Scotland to India. During the voyage 230 members of the regiment died, victims of scurvy. This was due largely to the obstinacy of the government in requiring the consumption of citrus fruit by those on long sea voyages. This requirement would have prevented the Highlanders from contracting the disease. For reasons never clearly stated, and although the citrus fruit was available, the medical officers onboard these ships never initiated the procedure. This is even more puzzling, given that the treatment had been discovered by James Lind over sixty years earlier.[218] Although this shows that the lack of timely implementation was not unusual, in the case of quinine it served to bury Baikie's achievements deeper and to become an additional factor in eliminating recognition of his deeds.

Although Baikie appears to be the first to validate Bryson's theory, he was not the only explorer of the period to utilize quinine as a preventative. The careers of John Kirk and David Livingstone had much in common with that of Baikie. They spent much of their lives in Africa and their time in East Africa coincided with Baikie's second voyage on the Niger River. They also were Scottish physicians who elected to work as naturalists and explorers rather than to practice medicine. Kirk even attended the University of Edinburgh and studied under many of the professors and lecturers who had taught Baikie. They were each keenly aware of the dangers posed by malaria on their travels in Africa. There is also evidence that Kirk and Livingstone had studied Baikie's narrative of his first voyage and his successful use of quinine as a prophylactic.[219]

However, the matter of the optimal dosage to be administered was encountered by all three explorers and was a key factor in the reluctance of medical schools to recommend using the drug as a preventative. A few years after Baikie's second voyage, David Livingstone wrote about his study of Baikie's two voyages and the preventative use of quinine. He also alluded to the reasons the medical community had not given its full endorsement to the procedure and would not for another twenty years. "To be effective, quinine [has] to be given daily in controlled amounts. Too much, and one could get a ringing in the ears, temporary deafness, and blindness called cinchonism. Too little, and quinine [ceases] to be an effective preventative."[220]

On his second voyage, it appears that Baikie stopped using quinine too soon. If he had adjusted the dosage, as Kirk and Livingstone had done, perhaps he would have prevented the deaths on the second voyage. If he had completed two major expeditions without the loss of life, his theory would have been validated and his name might have been inscribed in medical history.

Throughout his preparation at the University of Edinburgh, Baikie had been trained as a general practitioner, physician, surgeon, and apothecary conflated into one. He had been trained in clinical diagnosis to look at individual objects or parts of nature as components of an overall system. His ability to see the whole, in addition to just the parts, certainly influenced his approach to people and both his personal and professional relationships.

Constantly curious and observant, Baikie's interest in health went beyond his concern about the "fever." Throughout his time in Africa, he sought ways to remain fit and healthy in such an inhospitable climate. His tests and observations resulted in recommendations for the overall health of those placed in his charge and he developed an inclusive approach to spend long periods of time successfully in the African interior. He presented his findings in a report to the Foreign Office, which they received, ironically, shortly after his death:

1. At first do not change suddenly from European food to native food.

2. Sleep under cover and if possible, under a net curtain which not only will keep out small insects, but also miasma inhalations.

3. Do not be afraid of the sun but try to get gradually used to it.

4. Avoid unnecessary exposure to wet, and do not sit in wet clothes.

5. Avoid the plan recommended by some missionaries of shutting up the houses at night.

6. After a time, begin to use native food. Make sure it is fully cooked and that the diet includes more vegetables.

7. Avoid damp places and take warm drinks early in the morning.

8. When traveling, the head should be well-covered against the sun.

9. The first symptoms of fever will yield to the proper use of quinine. Dysentery is much more dangerous but should be left unattended to except for altering the diet.

10. Avoid the use of alcoholic drinks.[221]

Here, Baikie has returned to his recommendation of quinine as a treatment and no longer as a preventative. And miasma is still mentioned as the probable cause of malaria, as it would be forty years before it was discovered that "the small insects" and not the mist caused the disease. But these rules are quite remarkable in that they closely parallel the advice a physician would give to those traveling in tropical Africa today.

WRITER AND NATURALIST

If he is not remembered as an explorer, an advocate of fair trade or the medical practitioner who proved the benefits of quinine, Baikie should surely be remembered for his published works. His summary of his first voyage and his travel to Kano were articulate and informative, even to a modern audience. His numerous translations and documentations of newly discovered African languages and dialects were groundbreaking. He published widely on a variety of topics; translated books into Hausa, Arabic, Igbo and Yoruba languages; and compiled vocabularies of over fifty indigenous languages and dialects. His work as a naturalist was praised by both Hooker and Darwin, and his collections and publications were housed in museums and botanical gardens throughout England and Scotland.

His *Narrative* was very well received, as were many of his other essays and journals, but following his death, those responsible for editing and publishing his writings were either unable or not interested in doing so. Although Glover wrote an extensive summary of Baikie's second expedition, Baikie himself had nothing published about the voyage. A possible explanation for this absence of work about his time marooned on the river, and his subsequent stint at Lokoja, can be found in a passage from a book written by William Marwick:

> The notes which he kept came to this country, but he [Baikie] had written them so as only to be deciphered by himself. They were placed in the hands of the late Ferdinand Fitzgerald, who the writer knew very well. He spent much time endeavoring to arrange and decipher them but could not succeed in do so in such a manner that would have enabled him to produce anything worth publishing.[222]

After Fitzgerald had worked on them, Baikie's diary and papers were then entrusted to John Kirk. This colleague of Livingstone had returned from Africa and was serving as an administrator in the Foreign Office. Kirk, too, could "not extract anything of general interest" from the diary about the second voyage. More ironically, he also rejected what is surely Baikie's most significant discovery, his proof of the efficacy of quinine. "Whatever its benefits as a cure," said Kirk, "it seemed valueless as a preventative." Later historians of British policy suggested that "the demonstration of the prophylactic use of quinine made the Niger rather than the Sahara the obvious route to the Sudan."[223] But this was years after Kirk's official report had been accepted by the Foreign Office.

The Royal Geographical Society had been a sponsor of both expeditions and continued to publish Baikie's writings up to, and even after the time of his death. But, at this time, the Society was going through significant transitions. Baikie's time in Africa coincided with the granting of their Royal Charter along with their change in name and location. Following his initial voyage, Baikie was asked by the Society to make only one presentation to their membership and was never scheduled to make a public presentation. As Lloyd writes, the explorers of West Africa "lived at a time [when they] were seldom celebrated in such a manner." The Royal Geographical Society would be at the height of its influence in the 1880s; but by that time Baikie and his work were already in the process of fading into history.

A MAN OF THE SCOTTISH ENLIGHTENMENT

Perhaps part of the puzzle in understanding Baikie's obscure place in history also lies in the character of the man himself. Sir Walter Scott speaks of "the national disposition to wandering and adventure," which combined with the Scottish Enlightenment's advanced educational preparation, "conduced to lead the Scots abroad." Baikie clearly had a powerful wanderlust and it could be that his traveling to, and his personal experiences in these far-off lands mattered as much to him as his official reasons for being there.

He was selected by Macgregor Laird to open the center of West Africa to trade. Although he was successful in commerce he was concerned as much with the relationships he created as the profits earned. He developed a genuine affinity for his adopted people and land. He was a representative of the British government, but against all norms he adapted to the culture rather than expecting those with whom he lived to change. Through the dispatches, the government had shown a repeated lack of interest in treating the Africans as equals. He saw himself as a friend to Africa when others approached the continent with a need to modify what they had found.

He understood his task as helping to secure for Britain a commanding position in Africa, but on his terms. When he was rescued following his year of isolation on the river, although he was extremely ill he refused to return to England. He felt a responsibility to complete the charge given to him and remained on the Niger because he believed his work was not yet complete. Five years later, when Lokoja was successful, Baikie was afraid to leave what he had built through his personal efforts and beliefs. He thought there was no one capable of carrying on his work and simply ignored the Foreign Office's order to leave the river unless they provided a suitable replacement. His fears proved well grounded, as those who followed him at Lokoja failed, and soon the consulate was to be closed. But at last, faced with the failing health of his father and after already being absent when his mother died, he agreed to return to England.

But he never arrived. When finally beginning his journey home he died before he or others could write his story. He went about his daily tasks with purpose and direction, never feeling the need to be the center of attention or the recipient of recognition. To him, obtaining the defined goal was more important than the fact that he was the one who had accomplished it. His successes among the African people were built so much upon the power of his unique personality and approach that they could not be sustained after his death. Baikie achieved greatness in exploration, trade, natural science, linguistics, and medicine. He laid the cornerstone that others would build upon. But it was those individuals who would ultimately be remembered. To most of the world, William Balfour Baikie would be forgotten.

Baikie was asked upon his return from this first voyage, what his expedition had accomplished and what future good could he anticipate coming from his efforts. He responded:

> Our voyage had points of interest for the utilitarian, the commercial man, the man of science and the philanthropist. We have discovered a navigable river, conducting us into the heart of a large continent, and by means of its branches we brought immediate contact with many thousands of miles of new country.
>
> We have met on friendly terms with numerous tribes, all endowed by nature with what I term the "commercial faculty," ready and anxious to trade with us. We can also indicate a most important outlet for home manufactures. Central Africa can absorb thousands of cargoes of soft goods, eagerly bartering the raw cotton, vegetable oil and ivory for our goods.
>
> To the man of science, we would enumerate the additions made to our geographic knowledge, to the extent of new country examined and laid down, to the survey of a new river and the determination of the erroneousness of the theory which derived the Benue from Lake Tsad. We would also allude to the new tribes discovered and the additions to ethnology, philology and natural history.
>
> To the medical philosopher we would mention the results of our experience of the climate, our options on the hitherto much dreaded "African fever," and the confirmation of the views of those who have recommended to prevent rather than to cure by employing a rational instead of an empirical treatment.
>
> Finally, we would address the philanthropist by telling him of multitudes of human beings to whom he might well turn this attention. They are organized like ourselves, have similar affections and desires, but that, unlike the inhabitants of our happier clime, they have been for ages a prey to the strong. That naturally they are mild and friendly, apt to learn and desirous of being taught, ready to receive first impressions, whether of good or of evil.[224]

His achievements were so diverse and so numerous that there is not one aspect of his life that emerges above the others. There is no seminal event that one can point to and say, "Ah that was Baikie." A quirk of timing prevented his contact with Heinrich Barth, and in fact

Barth was safely on the coast before Baikie left the river. Because of this issue, the attempted meeting never happened, but this singular event could have been of such magnitude that William Balfour Baikie's place in history could have been solidified forever. If the two of them had met up as part of that initial voyage, the meeting would have predated Livingstone's and Stanley's meeting by fifteen years; and Ernest Marwick speculates that perhaps William Balfour Baikie would have become a household name rather than Stanley.

Four likely reasons why Baikie has been largely forgotten by science and by history can be identified: happenstance, professional backstabbing, his enlightened approach to Africa, and his personal stubbornness. First, Baikie was plagued with misfortune. Had he located and rescued Barth on his initial voyage as planned, he would have become instantly famous. After a lifetime of work in Africa, Stanley is remembered for one line, "Dr. Livingstone, I presume." Yet, through a series of circumstances beyond Baikie's control, he did not locate Barth and no memorable rescue took place.

Misfortune continued. After his initial voyage, Baikie returned to England within a few weeks of the nominations for the Royal Geographical Awards and timing again played a critical role. Like many major awards, to receive these honors required some concentrated and focused marketing on the part of the individual nominee. Unfortunately, there was no time for Baikie or his supporters to publicize his accomplishments sufficiently in order to assure his achievements were given adequate consideration. Added to this, he was nominated during a year that included the largest number of recommendations made in that decade. Had fate allowed him to return even one month later, he would have been among the following year's selections. There would have been time to schedule writing and speaking engagements and his accomplishments would have been apparent to the judges, his peers, and the public at large. His reputation would have been widely recognized and the accomplishments of his competitors would have weighed far less. This accident of timing played a major role in the outcome of the awards.

On Baikie's second voyage bad luck struck again. He suffered a shipwreck and was marooned on the river for over a year. The *Dayspring* was lost not because of Baikie's poor leadership but because of an error in judgment on the part of Lieutenant Glover. Similarly, Baikie remained stranded for over a year only because the rescue plan that had been put in place at the beginning of the voyage was entirely mismanaged by others. In a world of fierce competition and partisan politics, it did not matter that Baikie was not personally responsible for the disaster. As the head of the expedition the responsibility fell on him.

Unlike the first expedition, the second saw a significant loss of life due to malaria. Baikie was among the most affected by the disease, suffering severely from the disease he had fought so diligently to defeat. According to the records provided by Glover, several crew members were already ill ahead of their arrival at Fernando Po. Perhaps some of these men were already beyond recovery—the facts will never fully be known. Although the loss of life to fever on this voyage was minimal compared with earlier expeditions, the fact that people died while under Baikie's preventive treatment seems to have led him to lose faith in his earlier research and writing. Records of the day do not reveal exactly why he made this choice, but Baikie's own struggles with fever could have impaired his judgment. It is now known that the potency of quinine crystals varies greatly. Baikie had established a dosage based on a specific number of grains of quinine. On the first voyage out, and with a different set of variables, this approach was successful. On the second voyage, Baikie maintained his original, and successful regimen. But this time people died. His adherence to what he once believed to be a life-saving routine created an additional disaster in this ill-fated journey.

Had Baikie not lost all faith in the effectiveness of quinine as a preventative and treatment for malaria and returned home for a second time without a loss of life, perhaps his medical legacy would have placed his name in the history books alongside Walter Reed and his work on yellow fever. Dr. Reed has a research medal named for him and hospitals and research institutions were named in his honor. Reed's image appears on a postage stamp and a tropical medicine course bears his name. When the London School of Hygiene and Tropical Medicine was constructed in 1926, twenty-three names of tropical medicine pioneers were chosen to be listed on the school building in Keppel Street. Reed, the American, was included. Baikie was not.

Second, Baikie was undermined again and again by those in whom he had placed his confidence and trust, creating another factor that kept his story suppressed. Rear-Admiral Fredrick Beechey was a very powerful and influential man in England's relatively small world of global exploration. As the newly named president of the Royal Geographical Society, he was openly critical of Baikie and went so far as to accuse him of taking credit for accomplishments that belonged to others. Colleague and physician Hutchinson was a man in whom Baikie had placed his confidence and whom he had trusted enough to promote to surgeon. Hutchinson's response to this trust was to attempt to assume all credit for Baikie's research and findings related to quinine. Fellow crew member May was a capable and loyal associate in the first voyage and a man Baikie clearly held in high esteem. Ironically, he became one of the greatest obstacles to the success of the second expedition and was ultimately stricken from the expedition's rolls. Such an action is harsh and reflects May's unusually destructive behavior on the second voyage. Finally, the captains placed in charge of the two ships, and the doctors placed in charge of the medical needs of the crew, were all shown to be extraordinarily inept and were reprimanded upon their return to England. The captains and the doctors were assigned to the ships by others, not chosen by Baikie; leaving him no option but to work around their failings. And yet, as leader of the expedition, Baikie would receive all credit or all blame. Once again, it was blame.

In the year following Baikie's and Barth's return to England, Barth was once again nominated by the Society for one of their two top awards. By the time Barth received his medal, Baikie had been publicly challenged on his charting of the Benue, his assumption of command, and his discoveries on the use of quinine. The Society clearly believed that Baikie was too much of a liability to give him public recognition through one of their prestigious awards. Recipients of such honors reflected upon the Royal Geographic Society, which was too rooted in tradition and too closely aligned with the British government to take a chance on honoring someone who was being challenged on so many fronts. The promised presentations to the public and additional sponsored presentations for the Society's members never materialized. Although Baikie's future travels and publications were sponsored by the

Society, he was never again to be nominated for one of their awards. Baikie had been selected to serve the Society's needs yet was never to be recognized by it.

Third, Baikie's deeds have failed to be chronicled simply because he promoted a philosophy of life and regard for fellow humans that was out of step with his contemporaries, a man out of time. Baikie was a man of peace and compassion. He was not a missionary preaching peace but a unique officer of the Royal Navy who elected to use respect and empathy, commonsense and kindness—not force.

Baikie engaged Africans as peers and as friends. He adapted to the local culture, language, and dress. While other Europeans were still active in the slave trade, Baikie was quietly buying and then freeing slaves, providing them with opportunities for education and jobs and helping to ensure their future. Baikie was respected by the locals for his strength of character and for his understanding of Africans and their beliefs. This deep respect was not shared by his English counterparts.

Although slavery had ended fifty years earlier, England maintained a philosophy of racial segregation. The government, as well as the general population, believed that areas of cultural, political, social, and ideological objectives could be accomplished only through segregation of the races. England would "civilize" Africa through Christianity and commerce. In order to do this, traditional African economies needed to be completely uprooted and non-Christian religions replaced. Success demanded that those in command must maintain their English dress, language, and customs. But most of all, they must maintain their distance from other racial groups.

Baikie had succeeded, but in doing so had also refuted every element of his government's philosophy and approach to Africa. The British powers had reluctantly made him their consul, but only after trying to recall him. They relented only when they realized the massive commercial success he had created could move forward only under his direction and his approach. The government was willing to ignore his aberrant behavior, realizing it was one man's actions and affected only a relatively small area in a remote part of Africa. But to praise his efforts as the new, effective approach to African trade would have taken them in a commercial and, more importantly, social direction they did not believe and in which they were consequently unwilling to go.

Over 150 years after his death, Baikie is still a famous figure in central Nigeria. At the confluence of the Niger and Benue rivers, his name is still given to any white visitor who enters the area. It is logical that he is remembered where his approach was embraced by the inhabitants. It is also understandable that he is forgotten in England where his methods were the antithesis of the culture and beliefs of the government. Baikie lived safely and happily among his African colleagues, earning the admiration and respect of those people, but never of his own countrymen.

Finally, Baikie was a proud man and a man of strong convictions. In many ways these could be considered his most laudable qualities. However, he sometimes took these characteristics to extremes where pride and conviction could be viewed by others as obstinance and stubbornness. The very qualities that allowed him to succeed at Lokoja certainly contributed to his ultimate rejection in England.

Many great men and women have been called stubborn and inflexible. Livingstone certainly fits this description; however, he returned to England regularly, acknowledging the political games that must be played in order to secure continued support and to build a legacy. While at home, Livingstone published widely in a variety of journals and made dozens of public presentations. He then returned to Africa, where his exploits continued to be followed closely by those at home. Baikie's decision not to return to England following his second voyage kept him out of the public's eye. In addition, it also meant that nothing he wrote during the last seven years of his life would be published until long after his death. He had refused to submit his writings to the government because they failed to share his views. Baikie's plan was to complete his writings himself and the submit them when he saw a change of attitude in London. His early death ensured this would never happen and his story once more was pushed further into the background.

Any, or all, of these explanations could be factors in Baikie's general absence from African history. Authors in the late nineteenth century routinely wrote in a manner that appealed to the general public. Treating Africans as equals, living among them, and adopting their

Figure 15.1 Baikie's obituary photo © Orkney Library and Archive

customs would never gain traction, and no author at the time would ever be tempted to write about his life. Only Baikie could have successfully told his story, and he died before he could tell the world of his Africa.

William Balfour Baikie ended the narrative of his first voyage with a challenge to his countrymen. He challenged them to follow up on the good work that had been started. He wrote of the eradication of slavery, the spread of Christianity, and the benefits that could be acquired through trading European manufactured goods for the nearly endless supply of African commodities. But these were, in his mind, merely by-products of his overall message and lesson. Baikie wrote that England needed to pursue these goals as a "labour of love and aim at acquiring and retaining the glorious title of Friend of Africa," and "if all aspects of a society were treated equally, the result would be the best that the participants could have hoped for."[225] [226][227] Had the world listened, the course of African history might have been very different and William Balfour Baikie's role within its history honored and recognized (Figure 15.1).

EPILOGUE

In the very early days of my research, when my wife and I traveled to Kirkwall in the Orkney Islands, I spent a great deal of time at the building in the center of the town that had been the family home of the Baikies of Tankerness. Tankerness House, as I described earlier, is now a museum dedicated to the history of the Orkney Islands and has one room that is dedicated to the Baikie family. Among other objects, it contains a few items that had been collected along the Niger and Benue rivers by William Balfour Baikie during his initial voyage.

This seemed plausible as he had brought back cases of collected items to be catalogued while he was working at Royal Hospital Haslar. It also made sense that he would have brought a few of these items to be distributed to family members and friends before leaving on his second voyage. The objects collected by the museum had been made by people Baikie had encountered during the voyage and were generally made of wood, stone, hide, or bone. The problem the museum had encountered is that most of the items had never been identified. It was clear where they had come from, but not clear what the items were nor their intended use.

Although no expert in this area, I agreed to spend an afternoon looking at the items and to provide information on those I could identify. It also gave me an extended period to talk one-on-one with the museum staff member who seemed well informed about Baikie and his family. During that afternoon, over twenty years ago, I asked if the Baikie family had any descendants in the region with whom I could discuss their family's history. Bryce Wilson reports that there is still a single relative, related to William, living in Kirkwall. As William's brother and one sister had no children, this relative would have to be descended from Katherine's daughter or from the four children fathered by the half-brother, Samuel. But my museum colleague could not provide any information about this local resident.

I was told that Robert, the last direct descendent of the owners of Tankerness House, had left Kirkwall years earlier and had settled in Africa. He had returned as the twelfth Laird of Tankerness to complete the sale of the estate years earlier. This would seem to be borne out by Hugh Marwick in *The Baikies of Tankerness* who describes a "Mr. Robert Baikie, the last Baikie of Tankerness, [who is] now a resident in Rhodesia." As we continued our discussion

153

of the Baikie family, almost as an afterthought, I was told that a young man had visited the library, church, and museum a year before my visit. He had claimed to be a direct descendent of the explorer and it was verified that his surname was Baikie. Also, he was a black man from Nigeria.

As a man who enjoys irony I am pleased with this final discovery and I think Baikie would have enjoyed it as well. William Balfour Baikie spent his brief life in the service of a people he came to value, respect and even to love. Against the backdrop of his own culture still intent on enslavement, exploitation, Christian conversion, and military conquest, he stood alone. He was viewed by his government as uncooperative.

I find great satisfaction in the way his story has ended. Baikie's extended Orkney family, like England's early reasons for being in Africa, has almost ceased to exist. Scottish relatives of the senior side of the family had relocated to Africa in the mid-twentieth century. Most of the descendants who still carry the family name are Africans, descended from that earlier union between William Balfour Baikie and his African partner, or from the countless slaves whom he personally freed and who then took his name.

Baikie was a true friend of Africa and a champion for the people of the Niger. He was also willing to give his life for his beliefs. Now, 150 years later, if you are white and travel along the Niger River you will be called "Baikie." I think he would have liked that as well.

BIBLIOGRAPHY

Ajayi, F. J. A. (1965) *Christian Missions in Nigeria 1841–1891*. Evanston, IL:
Northwestern University Press, 1965.

Anderson, T. (1859) *The Edinburgh New Philosophical Journal*. Edinburgh: Adam Charles Black, 1959.

Augustin, A. (2002) *Meet You at the Paddington* New York: Hilton International, 2002.

Baikie, W. B. (2011) *A List of Books and Manuscripts Relating to Orkney and Zetland*, London: Historical Print Editions,
    1966.

Baikie, W. B. (1966) *Narrative of an Exploring Voyage up the Rivers Kwora and Binue in 1854*. London: Frank Cass, 1966.

Baikie, W. B. (2009) *Observations on the Hausa and Fulfulde Languages (1861)*,
London: Kessinger Publishing LLC, 2009.

Baikie, W. B. (1862) *The African Repository, Vol XXXVIII*. Washington, DC: William H. Moore, 1862.

Baikie, W. B and Heddle, R. *Historia Naturalis Orcadensis. Zoology. Part 1. Being a catalogue of the mammalia and birds
    hitherto observed in the Orkney Islands*, Edinburgh: J & W Paterson, 1948.

Baikie, W. B. and Kirk, J. (1867) "Notes from a Journey from Bida in Nupe, to Kano in Haussa, Performed in 1862."
    Journal of the Royal Geographical Society of London, Vol. 37, 82–108, 1967.

Birbeck, E. *A Visit to Haslar, 1916.* Kew Printing, Gosport, 2004.

Birbeck, E. and Holcroft, A. (2004) *A Historical Guide to the Royal Hospital Haslar*. Gosport: Royal Hospital Haslar, 2004.

Boyles, D. *African Lives.* New York: Ballantine Books, 1988.

Brown, J. F. (2003) "Geology and Landscape." In *The Orkney Book,* edited by Donald Osmand, Edinburgh: Birlinn, 2003.

Bryson, A. (2007) "An Account of the Origin, Spread & Decline of the Epidemic Fevers of Sierra Leone." Portsmouth:
    Royal Naval Museum, 2007.

Burns, A. (1963) *History of Nigeria.* London: George Allen and Unwin, 1963.

Carland, J. M. *The Colonial Office and Nigeria, 1898–1914.* Stanford: Hoover Institution Press, 1985.

Cook, A. N. (1943) *British Enterprise in Nigeria.* Philadelphia, PA: University of Pennsylvania Press, 1943.

Crawford, B. E. (2003) "Orkney in the Middle Ages." In *The Orkney Book,* edited by Donald Osmand, Edinburgh:
    Birlinn, 2003.

Crowder, M. (1978) *The Story of Nigeria.* London: Faber and Faber, 1978.

Crowther, S. (1855) *Journal of an Expedition up the Niger and Tshadda Rivers in 1854, Undertaken by Macgregor Laird in
    Connection with the British Government.* London: Church Missionary House, 1855.

Cunningham, P. (1850) *Handbook of London: Past and Present.* London: J. Murray. 1850.

Darwin Correspondence Project (1858) "Baikie, W. B. to Darwin, C. R." Cambridge: Cambridge University Library,
    Letter number 2214, 1858.

Davidson, B. *The African Past: Chronicles from Antiquity to Modern Times.* London: Longmans, Green and Co., 1964.

Davidson, B. *The African Slave Trade.* Boston, MA: Brown, Little and Company, 1980.

de Gramont, S. (1976) *The Strong Brown God: The Story of the Niger River*, Boston, MA: Houghton Mifflin, 1976.

Dike, K. O. (1956) *Trade and Politics in the Niger Delta 1830–1885*, Oxford: Clarendon Press, 1956.

Evans, I. L. (1929) *The British in Tropical Africa: An Historical Outline*. Cambridge: Cambridge University Press, 1929.

Fage, J. D. (1988) *A History of Africa.2nd edn:* London: Unwin Hyman, 1988.

Falola, T. and Heaton, M. M. (2008) *A History of Nigeria*. Cambridge:Cambridge University Press, 2008.

Fisher, C. T. (2004) "Fraser, Louis 1831–1866." *Oxford Dictionary of National Biography*. Oxford: Oxford University Press, 2004.

Flett, B. (2003) "Orcadian Explorer Embarked on an African Voyage of Discovery," *The Orcadian*, January 2, 2003.

Geary, W. N. M. (1965) *Nigeria under English Rule*. New York: Barnes and Noble, 1965.

Giddings, R. (1996) *Imperial Echoes: Eye-Witness Accounts of Victoria's Little Wars*. London: Leo Cooper, 1996.

Grey, J. E. (1873) *Specimens of Shield Reptiles*. London: British Museum, 1873.

Grierson, E. (1972) *The Death of the Imperial Dream*. New York: Doubleday, 1972.

Hastings, A. C. G. (1926) *The Voyage of the* Dayspring. London: Bodley Head, 1926.

Hempelmann, E and Krafts, K. (2013) "Bad Air Amulets and Mosquitoes: 2000 Years of Changing Perspectives on Malaria." *Malaria Journal* 12, no. 1 (2013): 232.

Herman, A. (2001) *How the Scots Invented the Modern World*. New York: Broadway Books, 2001.

Hibbert, C. (1982) *Africa Explored: Europeans in the Dark Continent 1769–1889*. New York: W.W. Norton & Company, 1982.

Hochschild, A. (2005) *Bury the Chains*. Boston, MA: Houghton Mifflin, 2005.

Hodgkin, T. (1975) *Nigerian Perspectives*. London: Oxford University Press, 1975.

Hollett, D. (1995) *The Conquest of the Niger by Land and Sea*. Abergavenny: P.M. Heaton., 1995

Hooker, W. J. (1857) *Hooker's Journal of Botany and Kew Gardens Miscellany,1849–1857*. (9 volumes). London: Reeve, Benham, and Reeve, 1849–1857.

Hossack, B. H. (1900) *Kirkwall in the Orkneys*. Kirkwall: William Peace & Son, 1900.

Howard, C. (1951) *West African Explorers*. London: Oxford University Press, 1951.

Hakluyt, R. (1972) *Principal Navigations, Voyages, Traffics and Discoveries of the English Nation,* London: Penguin Classics, 1972.

Igwe, S. O. (1987) *Education in Eastern Nigeria 1847–1975*. Lagos: Lofs Printers, 1987.

Ikime, O. (1977) *The Fall of Nigeria: The British Conquest*. London: Heinemann, 1977.

Iliffe, J. (1993) *Africans: The History of a Continent*. Cambridge: Cambridge University Press, 1933.

Kolapo, F. J. (2001) "European Explorers and Aspects of 19th Century Nupe History." Transactions of the Historical Society of Ghana, New Series, 4–5 (2001): 105–122.

Lepore, J. (2018) *These Truths*. New York: W. W. Norton, 2018.

Leyburn, J. G. (1962) *The Scotch-Irish: A Social History*. Chapel Hill, NC: University of North Carolina Press, 1962.

Liebowitz, D. (1999) *The Physician and the Slave Trade: The Livingstone Expeditions, and the Crusade Against Slavery in East Africa*. New York: W. H. Freeman, 1999.

Lloyd, C. (1973) *The Search for the Niger*. London: Collins, 1973.

Lockhart, B. J. and Lovejoy, P. (2005) *Hugh Clapperton into the Interior of Africa*. Leiden: Brill, 2005.

Charles Augustus Ludwig, C. A. (1786) *Grammar of the Fulde language,* London: Church Missionary House, 1876.

Mannix, D. P. and Cowley, M. (1962) *Black Cargoes: A History of the Atlantic Slave Trade 1518–1865*. London: Longmans, Green and Co., 1962.

Manson-Bahr, P. (1961) "The Malaria Story." *Proceedings of the Royal Society of Medicine* 54, no. 2(1961): 91–100.

Marwick, E. W. (1965) *William Balfour Baikie: Explorer of the Niger (A Centenary Survey)*. Kirkwall: Kirkwall Press.

Marwick, H. (1957) "The Baikies of Tankerness." *Orkney Miscellany* 4, 1957.

Mill, R. H. (1930) *The Record of the Royal Geographical Society 1830–1930*. Royal Geographical Society: London, 1930.

Mockler-Ferryman, A. F. (2017) *Up the Niger*. London: Forgotten Books, 2017.

Montague, A. (1864) "Certified Extract of Death." Freetown, Sierra Leone, 1864.

Bibliography

Orwell, G. (1934) *Burmese Days*. London: Harcourt Brace, 1934.

Moore, C. D. and Dunbar, A. (1968) *Africa Yesterday and Today*. New York: Bantam, 1968.

Muir, E. (1964) *An Autobiography*, London: Hogarth Press, 1964.

Murray, E. (2006) *Thomas Joseph Hutchinson 1820–1885, a Biography. Society for Latin American Studies*: online publication, 1 October. http://www.irlandeses.org.

Myers, C. H. (2016) "Steering the Seas of Reform: Education, Empirical Science, and Royal Naval Medicine, 1815–1860." Ph.D. Diss., University of Pittsburgh, 2016.

Obichere, B. I. (1982) *Studies in Southern Nigerian History*. London: Frank Cass, 1982.

Ohadike, D. C. (1994) *Anioma: A Social History of the Western Igbo People*. Athens, OH: Ohio University Press, 1994

Oliver, R. and Fage, J. D. (1990) *A Short History of Africa*. New York: Penguin, 1990.

Page, J. (1892) *Samuel Crowther: The Slave Boy Who Became Bishop of the Niger*. London: S. W. Partridge, 1892.

Painter, N. I. (2010) *The History of White People*. New York: W.W. Norton, 2010.

Park, M. (1864) *The Life and Times of Mungo Park in Africa*. Edinburgh: William P. Nimmo, 1864.

Parsons, T. H. (1999) *The British Imperial Century 1815–1914*. Lanham, MD: Rowman and Littlefield, 1999.

Porter, R. (1998) *The Greatest Benefit to Mankind: A Medical History of Humanity. New York*: W. W. Norton, 1998.

Poser, C. M. and Bruyn, G. W. (1999) *An Illustrated History of Malaria*. New York: Parthenon, 1999.

Records of the Admiralty, Naval Forces, Royal Marines, Coastguard, and Related Bodies. *H.M.S. Investigator*, 8 Sep 1864–6 Sep 1865. Admiralty, and Ministry of Defense, Navy Department: Ships' Logs Publication No. ADM 53/8615. Public Records Office, Kew, 1865.

Records of the Foreign Office (1855), Report. Dr. W. Balfour Baikie's Expedition up *Rivers Kwora (Kivora) and Tshadda (Niger)*. Confidential Print Numerical Series Publication No. FO 881/671. Public Records Office, Kew, 1855.

Records of the Foreign Office (1861) *Niger Expedition*. Political and Other Departments: Supplements to General Correspondence before 1906 Publication No. FO 97/433. Public Records Office, Kew, 1862.

Records of the Foreign Office (1862), Reports. Dr. Baikie's Niger Expedition. Map**.** Confidential Print Numerical Series Publication No. FO 881/1110. Public Records Office, Kew, 1862.

Records of the Foreign Office 1864, *Niger Expedition*, 1862–1864. Political and Other Departments: Supplements to General Correspondence before 1906 Publication No. FO 97/434. Public Records Office, Kew, 1864.

Records of the Foreign Office (1866) Journey of Dr. W. B. Baikie in 1862, from Bida in Nupe to Kano Hausa, Extracted from portions of *Dr. Baikie's Journals* in possession of F.O (Confidential Print Numerical Series Publication No. FO 881/1459). Public Record Office, Kew, 1866.

Ricks, T. E. (2018) *Churchill & Orwell,* New York: Penguin, 2018.

Shaw. N. (1858) *Proceedings of the Royal Geographical Society of London*, Vol. II, London: Royal Geographical Society, 1858.

Shaw, N. (1867) *Proceedings of the Royal Geographical Society of London*, Vol. XI, London: Royal Geographical Society, 1867.

Taylor, D. (1962) *The British in Africa*. New York: Roy, 1962.

Taylor, N. (1945) *Cinchona in Java: The Story of Quinine*. New York: Greenberg, 1945.

Temple, R. (1897) *Life of Sir John Hawley Glover*. London: Smith, Elder & Co, 1897.

The Orkney Herald "The Late Captain Baikie, R. N." Kirkwall: 1875.

Royal Anthropological Institute of Great Britain and Ireland. *Transactions of the Ethnological Society of London* (1867) London: John Murray, 1867.

Turner, A. L. (1937) *Story of a Great Hospital: The Royal Infirmary of Edinburgh 1729–1929*. Edinburgh: Oliver and Boyd, 1937.

Walker, F. D. (1930) *The Romance of the Black River*. London: Church Missionary Society, 1930.

Wenham, S. (2003) "Tankerness." In *The Orkney Book,* edited by Donald Osmand, Edinburgh: Birlinn, 2003.

Wickham-Jones, C. (2011) *Orkney: A Historical Guide*. Edinburgh: Birlinn, 2011**.**

The Orcadian. "William B. Baikie, M. D., R. N." Tuesday, January 17, 1865

Orwell, G. (1934) *Burmese Days*. London: Harcourt Brace, 1934.

Wilson, B. (2003) *Profit Not Loss: The Story of the Baikies of Tankerness*. Kirkwall: Orkney Heritage, 2003.

ENDNOTES

1 Boniface I. Obichere ed., *Studies in Southern Nigerian History* (London: Frank Cass. 1982), 23.

2 Thomas E. Ricks, *Churchill & Orwell* (New York: Penguin, 2018), 30–31.

3 Basil Davidson, *The African Past: Chronicles from Antiquity to Modern Times* (London: Longmans, Green, 1964), 3.

4 Dennis Boyles, *African Lives* (New York: Ballantine, 1988), 42.

5 Timothy H. Parsons, *The British Imperial Century 1815–1914: A World History Perspective* (Lanham, MA: Rowman and Littlefield, 1999), 72.

6 Robert Giddings, *Imperial Echoes: Eye-Witness Accounts of Victoria's Little Wars.* (London: Leo Cooper, 1996). xvii–xx.

7 Sir William N. M. Geary, *Nigeria under English Rule (1927)* (London: Frank Cass, 1965), 8.

8 Roland Oliver and J. D. Fage, *A Short History of Africa* (London: Penguin, 1988), 2.

9 Ernest Walker Marwick, *William Balfour Baikie: Explorer of the Niger; A Centenary Survey* (Kirkwell: Kirkwall Press, 1965), 7.

10 Records of the Admiralty, ADM 53/8615, *H.M.S. Investigator 8 Sep 1864–6 Sep 1865,* Ships' logs (Public Records Office, Kew, 1866).

11 Caroline Wickham-Jones, *Orkney: A Historical Guide* (Edinburgh: Birlinn, 2011). 109.

12 Barbara E. Crawford, "Orkney in the Middle Ages," in *The Orkney Book,* edited by Donald Osmand (Edinburgh: Birlinn, 2003), 78–79.

13 Arthur Herman, *How the Scots Invented the Modern World* (New York: Broadway Books, 2001), 363.

14 B. H. Hossack, *Kirkwall in the Orkneys.* (Kirkwell: William Peace & Son, 1900), 240–241.

15 Shiela Wenham, "Tankerness," in *The Orkney Book,* edited by Donald Osmand (Edinburgh: Birlinn, 2003), 197.

16 Hugh Marwick, *The Baikies of Tankerness* (Kirkwell: Orkney Miscellany. V.4, 1957), 33.

17 Bryce Wilson, *Profit Not Loss: The Story of the Baikies of Tankerness* (Kirkwell: Orkney Heritage, 2003), 67–78.

18 *The Orkney Herald,* "The Late Captain Baikie, R. N." (Wednesday, December 1, 1875).

19 Hossack, *Kirkwall in the Orkneys*, 330.

20 Ernest Walker Marwick, *William Balfour Baikie,* 7.

21 John Flett Brown, "Geology and Landscape" in Donald Osmand ed., *The Orkney Book.* (Edinburgh: Birlinn, 2003). 19.

22 William Balfour Baikie, *A List of Books and Manuscripts Relating to Orkney and Zetland,* (London, UK: British Library, Historical Print Editions, 2011).

23 William Balfour Baikie and Robert Heddle, Historia Naturalis Orcadensis. Zoology. Part 1. Being a catalogue of the mammalia and birds hitherto observed in the Orkney Islands, (Edinburgh, UK: J & W Paterson, 1948).

24 Edwin Muir, *An Autobiography,* (London, UK: Hogarth Press, 1964)

25 Turner, A. Logan, *Story of a Great Hospital: The Royal Infirmary of Edinburgh 1729–1929* (Edinburgh: Oliver and Boyd, 1937). 367.

26 Myers, C. H., "Steering the Seas of Reform: Education, Empirical Science, and Royal Naval Medicine, 1815–1860." (Ph.D. Diss., University of Pittsburgh, 2016).

27 J. D. Fage, *A History of Africa* (London: Unwin Hyman, 1988), 3.

28 David Hollett, *The Conquest of the Niger by Land and Sea* (Abergavenny, Gwent: P.M. Heaton Publishing, 1995), 11.

29 Sanche de Gramont, *The Strong Brown God: The Story of the Niger River* (Boston, MA: Houghton Mifflin, 1976), 26.

30 Basil Davidson, *The African Slave Trade* (Boston, MA: Brown, Little and Company, 1980), 23.

31 Hollett, *The Conquest of the Niger,* 15.

32 Christopher Lloyd, *The Search for the Niger* (London: Collins, 1973), 14.

33 Michael Crowder, *The Story of Nigeria* (London: Faber and Faber, 1978), p. 59.

34 John Iliffe, *Africans: The History of a Continent* (Cambridge: Cambridge University Press, 1993), 53–54.

35 Christopher Hibbert, *African Explored: Europeans in the Dark Continent 1769–1889* (London: W. W. Norton & Company, 1982), 15.

36 Toyin Falola and Matthew M. Heaton, *A History of Nigeria* (Cambridge: Cambridge University Press, 2008), 32–33.

37 Don Taylor, *The British in Africa* (New York: Roy Publishers, 1962), 15–16.

38 Davidson, *The African Slave Trade,* 26.

39 Oliver and Fage, *A Short History,* 99–105.

40 Lloyd, *Search for the Niger*, 14.

41 Clark D. Moore and Ann Dunbar eds., *Africa Yesterday and Today* (New York: Bantam Books, 1968), 107.

42 Davidson, *The African Slave Trade,* 11–12.

43 Edward Grierson, *The Death of the Imperial Dream* (New York: Doubleday, 1972), 85.

44 Davidson, *The African Past,* 8–29.

45 Richard Hakluyt, *Principal Navigations, Voyages, Traffics and Discoveries of the English Nation* (London, UK: Penguin Classics, 1972).

46 Lloyd, *Search for the Niger*, 15.

47 Jill Lepore, *These Truths* (New York: W. W. Norton, 2018), 45–46.

48 Adam Hochschild, *Bury the Chains* (Boston, MA: Houghton Mifflin, 2005), 54–55.

49 Ibid., 2–3.

50 Daniel P. Mannix and Malcolm Cowley, *Black Cargoes: A History of the Atlantic Slave Trade 1518–1865* (London: Longmans, Green, 1962), 74–75.

51 Ibid., 76–77.

52 A. C. G. Hastings, *The Voyage of* the Dayspring (London: John Lane the Bodley Head Limited, 1926), 75.

53 Hibbert, *African Explored*, 15–16.

54 C. Howard, ed., *West African Explorers* (London: Oxford University Press, 1951), 1.

55 Lloyd, *Search for the Niger*, 21–22.

56 William Balfour Baikie, *Narrative of an Exploring Voyage up the Rivers Kwora and Binue in 1854* (London: Frank Cass, 1966), 1–6.

57 Howard, *West African Explorers,* 4.

58 Hollett, *The Conquest of the Niger,* 11

59 Lloyd, *Search for the Niger*, 16–17.

60 Hibbert, *African Explored*, 18.

61 Howard, *West African Explorers,* 9.

62 James G. Leyburn, *The Scotch-Irish: A Social History* (Chapel Hill, NC: University of North Carolina Press, 1962), 45.

63 Ibid., 43.

64 Herman, *How the Scots Invented,* 25–26.

65 Leyburn, *The Scotch-Irish*, 71.

66 Herman, *How the Scots Invented*, 324–329.

67 Lloyd, *Search for the Niger*, 31–32.

68 Howard, *West African Explorers,* 84–85.

69 Hollett, *The Conquest of the Niger,* 42.

70 Mungo Park, *The Life and Times of Mungo Park in Africa* (Edinburgh: William P. Nimmo, 1864), 310–313.

71 Femi J. Kolapo,"European Explorers and Aspects of 19ᵗʰ Century Nupe History" (Acura: Historical Society of Ghana, Series 4–5, 2001), 108–109.

72 Howard, *West African Explorers,* 258–259.

73 Thomas Hodgkin, *Nigerian Perspectives* (London: Oxford University Press, 1975), 16–17.

74 Bruce J. Lockhart and Paul Lovejoy, *Hugh Clapperton into the Interior of Africa* (Leiden: Brill, 2005).

75 Parsons, *The British Imperial Century,* 64–65.

76 Oliver and Fage, *A Short History*, 135–137.

77 K. Onwuku Dike, *Trade and Politics in the Niger Delta 1830–1885* (Oxford: Clarendon Press, 1956), 168–170.

78 Hollett, *The Conquest of the Niger,* 176.

79 Hodgkin, *Nigerian Perspectives*, 17.

80 Hibbert, *African Explored*, 181–187.

81 Howard, *West African Explorers,* 387–388.

82 Hollett, *The Conquest of the Niger,* 12.

83 Lockhart and Lovejoy, *Hugh Clapperton.*

84 Hollett, *The Conquest of the Niger,* 157–158.

85 Howard, *West African Explorers,* 475.

86 Don C. Ohadike, *Anioma: A Social History of the Western Igbo People* (Athens, OH: Ohio University Press, 1994), 116–117.

87 Hollett, *The Conquest of the Niger,* 172.

88 S. Okoronkwo Igwe, *Education in Eastern Nigeria 1847–1975* (Lagos, NG: Lofs Printers, 1987), 46.

89 Hollett, *The Conquest of the Niger,* 179.

90 Hollett, *The Conquest of the Niger,* 185.

91 Baikie, *Narrative of an Exploring Voyage* 7.

92 Clemency Thorne *Fisher, "Fraser, Louis 1831–1866." Oxford Dictionary of National Biography (Oxford: Oxford University Press, 2004).*

93 Baikie, *Narrative of an Exploring Voyage* 8–10.

94 Baikie, *Narrative of an Exploring Voyage* 12.

95 Hollett, *The Conquest of the Niger,* 182–185.

96 Baikie, *Narrative of an Exploring Voyage* 18–20.

97 Baikie, *Narrative of an Exploring Voyage* 23.

98 Hollett, *The Conquest of the Niger,* 183.

99 Baikie, *Narrative of an Exploring Voyage* 27.

100 Brian Flett, "Orcadian Explorer Embarked on an African Voyage of Discovery," *The Orcadian* (Kirkwell: January 2, 2003), 1.

101 Sir Alan Burns, *History of Nigeria* (London: George Allen and Unwin,1963), 21.

102 Baikie, *Narrative of an Exploring Voyage* 29.

103 Baikie, *Narrative of an Exploring Voyage* 27–28.

104 Hollett, *The Conquest of the Niger,* 183–185.

105 P. Manson-Bahr, "The Malaria Story," *Proceedings of the Royal Society of Medicine* 54(2): 91–100, 1861.

106 E. Hempelmann and K. Krafts, "Bad Air Amulets and Mosquitoes: 2000 Years of Changing Perspectives on Malaria," *Malaria Journal* 12, no. 1 (2013).

107 Daniel Liebowitz, *The Physician and the Slave Trade: the Livingstone Expeditions, and the Crusade Against Slavery in*

East Africa (New York: W. H. Freeman, 1999), 44–45.

108 Hibbert, *African Explored*, 121.

109 C. M. Poser and G. W. Bruyn, *An Illustrated History of Malaria* (New York: Parthenon, 1999).

110 N. Taylor, *Cinchona in Java: The Story of Quinine* (New York: Greenberg, 1945).

111 Roy Porter, *The Greatest Benefit to Mankind: A Medical History of Humanity.* (New York: W. W. Norton, 1998), 9–11.

112 Alexander Bryson, *An Account of the Origin, Spread & Decline of the Epidemic Fevers of Sierra Leone* (Royal Naval Museum, 2007).

113 Baikie, *Narrative of an Exploring Voyage* 28.

114 Hodgkin, *Nigerian Perspectives*, 67

115 Igwe, *Education in Eastern Nigeria,* 47.

116 Samuel Crowther, *Journal of an Expedition up the Niger and Tshadda Rivers in 1854, Undertaken by Macgregor Laird in Connection with the British Government* (London: Church Missionary House, 1855), 167–170.

117 Baikie, *Narrative of an Exploring Voyage* 91–93.

118 Baikie, *Narrative of an Exploring Voyage* 99–110.

119 Baikie, *Narrative of an Exploring Voyage* 167–172.

120 Baikie, *Narrative of an Exploring Voyage* 195–200.

121 Baikie, *Narrative of an Exploring Voyage* 255–226.

122 Baikie, *Narrative of an Exploring Voyage* 265–266.

123 Hollett, *The Conquest of the Niger,* 194–195.

124 Hossack, *Kirkwall in the Orkneys,* 206.

125 Records of the Foreign Office (1855), Report. Dr. W. Balfour Baikie's Expedition up *Rivers Kwora (Kivora) and Tshadda (Niger).* Confidential Print Numerical Series Publication No. FO 881/671. Public Records Office, Kew, 1855.

126 J. F. Ade Ajayi, *Christian Missions in Nigeria 1841–1891* (Evanston, IL: Northwestern University Press, 1965), 209.

127 K. Onwuku Dike, *Trade and Politics in the Niger Delta 1830–1885* (Oxford: Clarendon Press, 1956), 168–170.

128 Andreas Augustin, *Meet You at the Paddington* (New York: Hilton International, 2002), 9–11.

129 Peter Cunningham, *Handbook of London: Past and Present* (London: J. Murray, 1850).

130 Eric Birbeck and A. Holcroft, *A Historical Guide to the Royal Hospital Haslar*, Gosport: Royal Hospital Haslar, 2004.

131 Ibid.

132 Ibid.

133 Ibid.

134 *Eric C. Birbeck, A Visit to Haslar (Gosport. Online publication of the Haslar Heritage Group, February 2014).*

135 R. H. Mill, *The Record of the Royal Geographical Society 1830–1930* (London: Royal Geographical Society, 1930).

136 Ibid.

137 Ibid.

138 Baikie, *Narrative of an Exploring Voyage*, ix.

139 Mill, *The Record of the Royal Geographical Society*.

140 Edmundo Murray, *Thomas Joseph Hutchinson 1820–1885, A Biography* (Society for Latin American Studies: online publication, 1 October 2006).

141 Lloyd, *Search for the Niger*, 197–198.

142 Baikie, *Narrative of an Exploring Voyage,* ix.

143 Hollett, *The Conquest of the Niger,* 204.

144 Lockhart and Lovejoy, *Hugh Clapperton.*

145 Records of the Foreign Office (1861) *Niger Expedition.* Political and Other Departments: Supplements to General Correspondence before 1906 Publication No. FO 97/433. Public Records Office, Kew, 1862.

146 Hollett, *The Conquest of the Niger,* 176.

147 Hooker, *Hooker's Journal*, 122.

148 Ajayi, *Christian Missions*, 43.

149 F. Deaville Walker, *The Romance of the Black River: The Story of the C.M.S. Nigeria Mission* (London: Church Missionary Society, 1930), Chapter 9.

150 Hastings, *The Voyage of the* Dayspring, 77–78.

151 Jessie Page, *Samuel Crowther: The Slave Boy Who Became Bishop of the Niger.* (London: S. W. Partridge, 1892), Chapter 8.

152 Records of the Foreign Office (1861) *Niger Expedition.* No. FO 97/433. Public Records Office, Kew, 1862.

153 Hastings, *The Voyage of the* Dayspring, 83–86.

154 Walker, *The Romance of the Black River,* Chapter 9.

155 Lockhart and Lovejoy, *Hugh Clapperton.*

156 Hastings, *The Voyage of the* Dayspring, 105–112.

157 Norton Shaw ed., *Proceedings of the Royal Geographical Society of London*, Vol. II (London: Royal Geographical Society, 1858).

158 Hollett, *The Conquest of the Niger,* 217.

159 Liebowitz, *The Physician and the Slave Trade*, 70–71.

160 Walker, *The Romance of the Black River,* Chapter 9.

161 Darwin Correspondence Project, "Baikie, W. B. to Darwin, C. R.," (Cambridge: Cambridge University Library, 1858). Letter Number 2214.

162 Ibid.

163 Thomas Anderson, ed., *The Edinburgh New Philosophical Journal* (Edinburgh: Adam Charles Black, 1859). 124–125.

164 Hastings, *The Voyage of the* Dayspring, 206–207.

165 Hastings, *The Voyage of the* Dayspring, 210.

166 Hastings, *The Voyage of the* Dayspring, 208–209.

167 Hastings, *The Voyage of the* Dayspring, 210.

168 Records of the Foreign Office (1861) *Niger Expedition.* No. FO 97/433. Public Records Office, Kew, 1862.

169 Obaro Ikime, *The Fall of Nigeria: The British Conquest* (London: Heinemann Publishing, 1977) 214.

170 Walker, *The Romance of the Black River*, Chapter 9.

171 Crowder, *The Story of Nigeria*, 131.

172 Norton Shaw ed., *Proceedings of the Royal Geographical Society of London*, Vol. XI (London: Royal Geographical Society, 1867).

173 Records of the Foreign Office (1861) *Niger Expedition.* No. FO 97/433. Public Records Office, Kew, 1862.

174 Arthur Norton Cook, *British Enterprise in Nigeria* (Philadelphia, PA: University of Pennsylvania Press, 1943), 44.

175 Augustus Ferryman Mockler-Ferryman, *Up the Niger* (London: Forgotten Books, 2017), 283–284.

176 Lloyd, *Search for the Niger*, 202.

177 Herman, *How the Scots Invented,* 354–355.

178 Parsons, *The British Imperial Century,* 67.

179 Sir Richard Temple, ed., *Life of Sir John Hawley Glover* (London: Smith, Elder & Co. 1897), 104.

180 Grierson, *The Death,* 84–86.

181 Parsons, *The British Imperial Century,* 126.

182 Lloyd, *Search for the Niger*, 200–201.

183 Nell Irvin Painter, *The History of White People* (New York: W.W. Norton & Company, 2010), 201–202.

184 William Balfour Baikie, *The African Repository, Vol XXXVIII* (Washington, DC: William H. Moore, 1862.)

185 Records of the Foreign Office (1862), Reports. Dr. Baikie's Niger Expedition. Map**.** Confidential Print Numerical Series Publication No. FO 881/1110. Public Records Office, Kew, 1862.

186 George Orwell, *Burmese Days* (London: Harcourt Brace, 1934), 118

187 Iliffe, *Africans: The History,* 149.

188 Burns, *History of Nigeria*, 140.

189 Ohadike, *Anioma*, 97.

190 Ibid., 104–113.

191 Lloyd, *Search for the Niger*, 203.

192 Records of the Foreign Office (1861) *Niger Expedition*. No. FO 97/433. Public Records Office, Kew, 1862.

193 Lloyd, *Search for the Niger*, 203

194 Crowder, *The Story of Nigeria,* 82–83.

195 Hodgkin, *Nigerian Perspectives*, 325–327.

196 Iliffe, *Africans: The History,* 170–172.

197 Records of the Foreign Office (1866) Journey of Dr. W. B. Baikie in 1862, from Bida in Nupe to Kano Hausa, Extracted from portions of *Dr. Baikie's Journals* in possession of F.O (Confidential Print Numerical Series Publication No. FO 881/1459). Public Record Office, Kew, 1866.

198 William Balfour Baikie and John Kirk, *Notes from a Journey from Bida in Nupe, to Kano in Haussa, Performed in 1862.* Journal of the Royal Geographical Society of London, 37 (1866): 82–108.

199 Hodgkin, *Nigerian Perspectives,* 322–323.

200 Ibid.

201 Hibbert, *African Explored*, 109.

202 Ibid.

203 Records of the Foreign Office (1862), Reports. Dr. Baikie's Niger Expedition. Map. No. FO 881/1110. Public Records Office, Kew, 1862.

204 Hollett, *The Conquest of the Niger,* 252.

205 William Balfour Baikie, *Observations on the Hausa and Fulfulde Languages (1861),* (London, UK: Kessinger Publishing LLC, 2009.

206 Charles Augustus Ludwig, *Grammar of the Fulde language,* (London, UK: Church Missionary House, 1876).

207 J. E. Grey, *Specimens of Shield Reptiles.* British Museum (London: Museum Trustees, 1873).

208 Records of the Foreign Office 1864, *Niger Expedition*, 1862–1864. Political and Other Departments: Supplements to General Correspondence before 1906 Publication No. FO 97/434. Public Records Office, Kew, 1864.

209 Royal Anthropological Institute of Great Britain and Ireland, *Transactions of the Ethnological Society of London* (London: John Murray, 1867), 84.

210 Records of the Admiralty. ADM 53/8615, *H.M.S. Investigator 8 Sep 1864.*

211 *The Orcadian.* "William B. Baikie, M. D., R. N.," Tuesday, January 17, 1865.

212 Montague, A. "Certified Extract of Death," (Freetown, Sierra Leone, 1864).

213 Shaw, *Proceedings of the Royal Geographical Society of London*, Vol. XI

214 Ifor L. Evans, *The British in Tropical Africa* (New York: Negro Universities Press, 1969), 134–135.

215 Lloyd, *The Search for the Niger*, 204–205.

216 John M. Carland (1985) *The Colonial Office and Nigeria, 1898–1914.* (Stanford, CA: Hoover Institution Press, 1985).

217 Porter, *The Greatest Benefit.*

218 Herman, *How the Scots Invented,* 149.

219 Liebowitz, *The Physician and the Slave Trade.*

220 Liebowitz, *The Physician and the Slave Trade,* 70–71.

221 Records of the Foreign Office 1864, *Niger Expedition*, 1862–1864. Publication No. FO 97/434. Public Records Office, Kew, 1864.

222 Marwick, *William Balfour Baikie,* 14.

223 Marwick, *William Balfour Baikie,* 15

224 Herman, *How the Scots Invented,* 149.

225 Baikie, *Narrative of an Exploring Voyage* 394–397.

226

227